London & North Eastern Railway 4-4-0 Tender Locomotives

Front cover photo:
D10 62656 *Sir Clement Royds* at Manchester Central, 1952. (Manchester Locomotive Society – MLS)

Back cover photo:
D16/3 62614, formerly 8783, one of two D16s painted in LNER apple green livery post the Second World War and maintaining the livery in the early days of nationalisation, seen here on shed, c1951. (Ian MacCabe colourisation)

London & North Eastern Railway 4-4-0 Tender Locomotives

Great Northern, Great Central, Great Eastern, Midland & Great Northern Joint Railway

DAVID MAIDMENT

AN IMPRINT OF PEN & SWORD BOOKS LTD.
YORKSHIRE – PHILADELPHIA

First published in Great Britain in 2024 by
Pen and Sword Transport
An imprint of Pen & Sword Books Ltd.
Yorkshire - Philadelphia

Copyright © David Maidment, 2024

ISBN 978 1 39903 680 1

The right of David Maidment to be identified as author of this work has been asserted by him in accordance with the Copyright, Designs and Patents Act 1988.

A CIP catalogue record for this book is available from the British Library.

All rights reserved. No part of this publication may be reproduced or transmitted in any form or by any means, electronic or mechanical, including photocopy, recording or any information storage and retrieval system, without the prior written permission of the publisher, nor by way of trade or otherwise shall it be lent, re-sold, hired out or otherwise circulated without the publisher's prior consent in any form of binding or cover other than that in which it is published and without a similar condition including this condition being imposed on the subsequent purchaser.

Typeset in Palatino by SJmagic DESIGN SERVICES, India.
Printed and bound by Printworks Global Ltd, London/Hong Kong.

Pen & Sword Books Ltd. incorporates the imprints of Pen & Sword Books: After the Battle, Archaeology, Atlas, Aviation, Battleground, Discovery, Family History, History, Maritime, Military, Politics, Select, Transport, True Crime, Fiction, Frontline Books, Leo Cooper, Praetorian Press, Seaforth Publishing, Wharncliffe and White Owl.

For a complete list of Pen & Sword titles please contact

PEN & SWORD BOOKS LIMITED
George House, Units 12 & 13, Beevor Street, Off Pontefract Road,
Barnsley, South Yorkshire, S71 1HN, England
E-mail: enquiries@pen-and-sword.co.uk
Website: www.pen-and-sword.co.uk

or

PEN AND SWORD BOOKS
1950 Lawrence Rd, Havertown, PA 19083, USA
E-mail: uspen-and-sword@casematepublishers.com
website: www.penandswordbooks.com

All David Maidment's royalties from this book will be donated to the Railway Children charity [reg. no. 1058991] [www.railwaychildren.org.uk]

Other books by David Maidment:
Novels (Religious historical fiction)
The Child Madonna, Melrose Books, 2009
The Missing Madonna, PublishNation, 2012
The Madonna and her Sons, PublishNation, 2015
The Reluctant Traitor, PublishNation, 2021

Novels (Railway fiction)
Lives on the Line, Max Books, 2013
Steamy Stories, PublishNation, 2021

Non-fiction (Railways)
The Toss of a Coin, PublishNation, 2014
A Privileged Journey, Pen & Sword, 2015
An Indian Summer of Steam, Pen & Sword, 2015
Great Western Eight-Coupled Heavy Freight Locomotives, Pen & Sword, 2015
Great Western Moguls and Prairies, Pen & Sword, 2016
Southern Urie and Maunsell 2-cylinder 4-6-0s, Pen & Sword, 2016
Great Western Small-Wheeled Double-Framed 4-4-0s, Pen & Sword, 2017
The Development of the German Pacific Locomotive, Pen & Sword, 2017
Great Western Large-Wheeled Double-Framed 4-4-0s, Pen & Sword, 2017
Great Western Counties, 4-4-0s, 4-4-2Ts & 4-6-0s, Pen & Sword, 2018
Southern Maunsell Moguls and Tank Engines, Pen & Sword, 2018
Southern Maunsell 4-4-0s, Pen & Sword, 2019
Great Western Granges, Pen & Sword, 2019
Cambrian Railways Gallery, Pen & Sword, 2019
Great Western Panniers, Pen & Sword, 2019
Great Western Kings, Pen & Sword, 2020
Great Western & Absorbed Railway 0-6-2Ts, Pen & Sword, 2020
Drummond's L&SWR Passenger & Mixed Traffic Locomotives, Pen & Sword 2020
Southern 0-6-0 Tender Locomotives, Pen & Sword, 2021
LNER 4-6-0 Locomotives, Pen & Sword, 2021
Midland & LMS 4-4-0s, Pen & Sword, 2021
Great Western Castle 4-6-0 Locomotives, 1923-1959, Pen & Sword, 2022
Great Western Castle 4-6-0 Locomotives, The Final Years 1960-1965, Pen & Sword, 2022
Great Western Castle 4-6-0 Locomotives, In the Preservation Era, Pen & Sword, 2023
Four-coupled Tank Locomotives, Built by the Great Western Railway, Pen & Sword, 2023
Four-coupled Tank Locomotives, Absorbed by the Great Western Railway, Pen & Sword, 2023
The Princess Coronation Pacific Locomotives, 1937-1956, Pen & Sword, 2023
The Princess Coronation Pacific Locomotives, The Final Years & Preservation, Pen & Sword, 2023
Great Western 0-6-0 tender locomotives, Pen & Sword, 2024

Non-fiction (Street Children)
The Other Railway Children, PublishNation, 2012
Nobody ever listened to me, PublishNation, 2012

CONTENTS

Preface & Acknowledgements 6

Introduction 7

Chapter 1   **The Engineers** 8
- H.A. Ivatt 8
- Charles Sacré 8
- Thomas Parker 8
- Henry Pollitt 9
- John G. Robinson 9
- James Holden 9
- Stephen D. Holden 10
- Alfred J. Hill 10
- Samuel Johnson 10

Chapter 2   **The Great Northern 4-4-0s** 11
- The D1, H.A. Ivatt, 1911 11
- The D2, H.A. Ivatt, 1898 18
- The D3 & D4, H.A. Ivatt, 1896 30

Chapter 3   **The Great Central 4-4-0s** 43
- The D5, Henry Pollitt, 1895 43
- The D6, Henry Pollitt, 1897 47
- The D7, Thomas Parker, 1887 56
- The D8, Thomas Parker, 1888 64
- The D9, J.G. Robinson, 1901 67
- The D10, J.G. Robinson, 1913 89
- The D11/1, J.G. Robinson, 1919 113
- The D11/2, J.G. Robinson/Nigel Gresley, 1924 142
- The D12, Charles Sacré, 1877 157

Chapter 4   **The Great Eastern 4-4-0s** 162
- The D13, James Holden, 1886 162
- The D14, James Holden, 1900 170
- The D15, James Holden, 1903 173
- The D16, Alfred Hill, 1923 179

Chapter 5   **The Midland & Great Northern Railway 4-4-0s** 207
- The 'A' Rebuilds, Beyer Peacock, 1882 207
- The D52, S.W. Johnson, 1894 209
- The D53 & D54, S.W. Johnson Rebuilds 212

Colour Section 217

Appendix 225

Bibliography 245

Index 246

PREFACE & ACKNOWLEDGEMENTS

I am following up my earlier Pen & Sword book on the LNER 4-6-0s with similar books on that company's 4-4-0 locomotives, the majority designed and constructed by the railways that became the constituent parts of the LNER in 1923. Only the Gresley D49s (the Shires and the Hunts) were designed and constructed in the LNER era, although the Great Central D11 (Directors) were constructed in 1924 for the LNER's Scottish Division. Because of the sheer number of classes to be covered I am writing two volumes for Pen & Sword, the first of which will cover the 4-4-0s of the Great Northern, Great Central and Great Eastern Railways, plus a short chapter on the Midland and Great Northern locomotives that passed to LNER stock in October 1936. As most of these locomotives were of, or similar to, the Derby Midland designs of Samuel Johnson, they are dealt with in greater depth in my earlier Pen & Sword book on the Midland and the London, Midland and Scottish (LMS) 4-4-0s. The 4-4-0s of the North Eastern, North British and Great North of Scotland Railways plus the Gresley D49 designs will be covered in a second Pen & Sword book that will be published later.

I acknowledge the considerable help of Paul Shackcloth of the Manchester Locomotive Society, who presides over its huge archive of photographs and allows me to include them for publication without any fee as the royalties from the book will be donated, as with my royalties from other books I have written, to the Railway Children charity (www.railwaychildren.org.uk) which I founded in 1995 and which supports street and runaway children around the railway and bus stations of India, East Africa and the United Kingdom. The archive contains so many photographs of LNER 4-4-0s that I have hardly needed to search elsewhere. I acknowledge a few from John Scott-Morgan's collection to fill the gaps plus a few early ones I took myself to illustrate my limited personal experience of living and working with these locomotives. Many of these photos, however, bear no details of the original photographer. I have endeavoured to trace the copyright holder where known, but if I have missed anyone, please contact the publisher and I will try to make amends. I also acknowledge access to the locomotive performance logs of the Railway Performance Society and their contributors and their records culled from early articles in the *Railway Magazine* by gurus such as Cecil J. Allen.

I thank in particular my editor, Carol Trow and Pen & Sword's Production Manager, Janet Brookes, for their very professional support and the whole of the Pen & Sword team, designers, marketing and production staff for the consistent quality of the books they produce on my behalf.

David Maidment
2024

INTRODUCTION

Passenger traffic in the late nineteenth century grew rapidly and the various railway companies, which had relied on single-wheelers for the majority of their existence, were being forced to look at ways of coping with heavier train loads. The famous Stirling and Johnson 4-2-2s and the Great Western (GW) Dean engines were still on main line work, but their loads were limited. The first 4-4-0s of any consequence appeared on the Midland in the late 1870s and a few classes appeared on various railways in the 1880s, but it was the 1890s when the 4-4-0 wheel arrangement flourished on all the main line railway companies of the pre-Grouping age. The first decade of the twentieth century saw the introduction of 4-4-0 classes that reigned on the main express services until the 4-6-0 or 4-6-2s became more widespread. The most famous and most competent were the GW 'Cities', the London and North Western Railway (LNWR) 'George Vs', the GC 'Directors', the North Eastern (NE) 'Rs', the Midland 'Compounds', the Caledonian 'Dunalastairs', the Great Eastern (GE) 'Claud Hamiltons', the South Eastern and Chatham Railway (SE&CR) 'Ds' and 'Es' and the London and South Western Railway (LS&WR) 'T9s'. Only on the GW and GE were they replaced on the best expresses by 4-6-0s within their first decade. The Great Central, Northern Eastern and North British produced 4-4-2s for express work in a similar timescale which shared top link work with the 4-4-0s. Only the Great Northern constructed the 4-4-2 'Atlantic' as its prime express locomotive in the first decade of the twentieth century, with its 4-4-0s taking mainly secondary roles.

The only completely new 4-4-0 designs produced by the 'Big Four' were the London and North Eastern Railway (LNER) D49 'Shire' and 'Hunt' classes built mainly for secondary work in the North East and Scotland and the Southern 'Schools' built initially for the Hastings route and then both Eastern Section mainlines to the Kent Coast via Chatham and Ashford. The 'Schools' were almost certainly the most powerful and competent of all the 4-4-0 British designs.

I have covered the GW 4-4-0s, the Midland 4-4-0s and the Southern 4-4-0s in previous Pen & Sword 'Locomotive Portfolios' and my book on the LNER 4-6-0s was published in 2021. I have now decided that it was time to tackle the LNER's 4-4-0 classes inherited from its constituent companies in a similar fashion and my two volumes will cover all classes designated D1 to D54 – all 4-4-0 tender engines apart from a few locomotives of class D50 and D51, 4-4-0 tank engines built by the North British Railway. This first volume will cover the design, construction, history, operation and performance of the Great Northern, LNER D1–D4, Great Central, D5–D12, Great Eastern, D13–D16, and the Midland and Great Northern D52–D54 engines of Midland ancestry which were utilised in East Anglia by the LNER after it took over that independent company in 1936. I have covered those Midland and Great Northern (M&GN) classes previously in my book on the Midland 4-4-0s, but for sake of completeness I am repeating them here, concentrating on their period in LNER ownership and operation, seeking out new and unpublished photographs to illustrate their work. The second volume will cover the 4-4-0 designs of the North Eastern Railway, the North British, the Great North of Scotland and the Gresley D49s plus the single Thompson D49 rebuild. I have endeavoured to find many previously unpublished photographs, both detailed portraits of especial interest for modellers and images of the classes in action to illustrate their operation. Information about the operation and particularly the performance of some of the less well-known classes has been hard to come by, but I have included whatever I could find from my researches.

Chapter 1
THE ENGINEERS

H.A. Ivatt

Henry Ivatt was born on 16 September 1851 at Wentworth in Cambridgeshire and was educated at Liverpool College. At the age of 17 he was apprenticed to John Ramsbottom at Crewe Works. He then acted as a locomotive fireman for six months and held various other posts including work in the Traffic Department culminating in appointment as Assistant Foreman at Stafford depot. He was appointed as Head of Holyhead Locomotive Depot in 1874 and in 1876 was made Head of the Chester Motive Power District. In 1877 he moved to Cork in Ireland as District Locomotive Superintendent of the Southern Division of the Great Southern & Western Railway. In 1882 he moved to Inchicore as Assistant Locomotive Engineer & Works Manager and in 1886 became Locomotive Engineer of the company.

In 1895 he returned to the mainland following Patrick Stirling as Locomotive Engineer of the Great Northern Railway (GNR), based at Doncaster. He was a popular manager with his staff and had a strong family life, fathering six children, although the eldest died in 1898. One of his sons, Henry George, followed in his father's footsteps, becoming Chief Mechanical Engineer of the LMS in 1946.

He became a member of the Institute of Mechanical Engineers in 1887, and a member of its Council between 1900 and 1922. He was elected Vice-President of the I.Mech.E in 1922-3, a role he held until his death, aged 72, at Haywards Heath in Sussex, on 25 October 1923.

Charles Sacré

Charles Reboul Sacré was born on 4 September 1831, one of thirteen children of John Joseph Berlot de Sacré, of French Huguenot descent. He was articled to Archibald Sturrock at the Great Northern Railway Works at Boston in 1846 and by 1853 had become Assistant Locomotive Superintendent at Peterborough. On 1 April 1859, he was appointed Chief Engineer and Locomotive Engineer of the Manchester, Sheffield and Lincolnshire Railway based at Gorton.

He was resident there for many years, popular with his staff, providing the railway with competent 0-6-0 goods engines as well as large diameter wheel 2-2-2s for express passenger services before the 4-4-0s described in this book. One of these, 434, was involved in the Penistone train accident of July 1884 which affected Sacré badly. The locomotive broke a crank axle and derailed but the tragedy was compounded by the fact that the train, without automatic brakes, ran away and caused the death of nineteen people including his friend Massey Bromley, Locomotive Engineer of the Great Eastern Railway. He felt guilty that he had given in to pressure from the company chairman, Edward Watkin, not to go to the expense of fitting the company trains with automatic vacuum brakes. The Inquiry cleared Sacré of any blame, but he chose to retire the following year although he was retained as a consultant. The outcome of the Penistone accident continued to prey on his mind, and he committed suicide by shooting himself on 3 August 1889.

He was a member of the Institute of Mechanical Engineers from 1859 and the Institute of Civil Engineers from 1867.

Thomas Parker

Thomas Parker was born in Ayrshire on 11 July 1829 and became an apprentice at the Caledonian Railway Works at Greenock. In 1858 he was appointed as the Carriage and Wagon Superintendent of the Manchester, Sheffield and Lincolnshire Railway at Gorton, and made a number of innovations in carriage design including some of the first dining car carriages in 1885. He was appointed to follow Charles Sacré when that gentleman resigned in 1885 and combined his C&W Engineer role with that of Locomotive Superintendent.

In 1891, he was the first British railway company engineer to incorporate the Belgian Belpaire firebox into his designs, although it had been used by the Beyer Peacock company for its export designs since 1872. In 1893 he implemented the provision of the continuous automatic brake system onto the company's rolling stock, an outcome of the Penistone accident referred to earlier. He retired in 1894 though remaining in the area and died at Gorton in 1903.

Henry Pollitt

Harry Pollitt as he was known was born at Ashton-under-Lyne on Boxing Day 1865. His father, Sir William Pollitt, became General Manager of the Manchester, Sheffield and Lincolnshire Railway in 1894 and Harry, who had been the Works Manager at Gorton, was appointed in the same year as Locomotive Engineer, following the retirement of Thomas Parker. His role changed to that of Locomotive and Marine Engineer as his role was enlarged to encompass the maintenance of the ferry services across the Humber.

In 1897 the Great Central Railway encompassed the Manchester, Sheffield and Lincolnshire Railway (MS&LR) and Harry Pollitt became its first Locomotive Engineer. He resigned, still aged only 35, in June 1900 and married an Australian woman, Mabel Amanda Alves, in 1901. He never worked again, presumably well endowed with the fortune of his father and died a year after his wife in Bournemouth on 23 January 1945.

John G. Robinson

John George Robinson was born in Newcastle-upon-Tyne on 30 July 1856. He was the second son of the locomotive engineer Matthew Robinson and his wife Jane. He was educated at Chester Grammar School and in 1872 was a GWR apprentice at Swindon, a pupil of Joseph Armstrong. In 1978 he was appointed as Assistant Engineer to his father at Bristol.

In 1884 he moved to Ireland as Assistant Carriage & Wagon Superintendent of the Waterford and Limerick Railway and the following year was appointed as Locomotive, Carriage and Wagon Superintendent. In 1900, following the sudden resignation of Harry Pollitt, he moved back to the mainland as Locomotive and Marine Superintendent of the Great Central Railway, his post retitled as Chief Mechanical Engineer in 1902. He held that post until the Grouping at the end of 1922 and was offered the similar post in the new LNER organisation, but being already 66 years of age, felt the post should go to a younger man and recommended Nigel Gresley. He was persuaded to remain for a while as an engineering consultant to the new company.

His design of the 2-8-0 freight locomotive had played a key role in military activities on the Continent in the First World War, and he was awarded the CBE in the 1920 Honours List. He died aged 87 on 7 December 1943.

James Holden

James Holden was born on 26 July 1837 in Whitstable in Kent and was apprenticed to his uncle, the North Eastern Locomotive Engineer Edward Fletcher, at Gateshead between 1852 and 1858. In 1865 he was appointed as manager of the Carriage and Wagon Department at Swindon Works and was assistant to William Dean. In 1885 he became Locomotive Superintendent of the Great Eastern Railway where he set about improving the efficiency of that railway which had difficult financial problems. He organised the compiling of a register of all the GER's locomotive and rolling stock assets with reliable and accessible information and began the standardisation of the locomotive fleet and its components.

He had an enquiring engineering mind and his catch phrase of 'Tell me this…' became well known as he went about his business. He was a Quaker which impacted on his attitude towards his staff and trade unions. He was somewhat paternalistic and believed it was his role as manager rather than that of the trade unions to provide for his men. He erected the first hostel for train crews in London for men stranded after running late.

He lived in Wanstead and during his time as Locomotive Superintendent reorganised Stratford Works, pioneered oil firing of many of his locomotives and introduced water scoops and troughs on the railway. He became a member of the Institute of Mechanical Engineers in 1886 and retired in 1908 handing over the reins to his son, Stephen. He died at Bath on 25 May 1925, aged 87.

Stephen D. Holden

Stephen Dewar Holden, the son of James, was born on 23 August 1870 at Saltney, Cheshire, while his father was working at Chester for the GWR. He was the family's third son and was educated privately and at University College, London. He left still aged only 16 and joined the Great Eastern Railway at Stratford where his father was now Locomotive Superintendent. He studied under his father in the Drawing Office and became a Running Inspector.

In 1892 he was appointed Suburban District Locomotive Superintendent, in 1894, District Locomotive Superintendent at Ipswich and in 1897 Divisional Locomotive Superintendent in London, where he ran the Locomotive Running Department and was Assistant to his father. He was appointed Locomotive Engineer of the GER in 1908 when his father retired, but resigned in 1912, aged only 42 and died in Rochester on 7 February 1918, seven years before his father.

Alfred J. Hill

Alfred Hill was born on 1 January 1862 in Peterborough and was educated at Waternewton Rectory in Northants. He was apprenticed aged 15 for six years from 1877 at Stratford Works and was awarded the Whitworth Scholarship in 1882. By 1891 he was Assistant Works Manager and was for twelve years a teacher at the Stratford Mechanics Institute. In 1899 he was appointed the Stratford Works Manager and in 1912 became the GER Locomotive Superintendent after the resignation of Stephen Holden. The post was retitled Chief Mechanical Engineer in 1915.

During the First World War he was Chairman of the Southern group of railways munitions committee working for the Ministry of Munitions and in 1917 went to the USA to improve the supply of materials for munitions. For his wartime service he was awarded the CBE.

He was a member of the Institute of Mechanical Engineers from 1901 and the Institute of Civil Engineers from 1910. He was President of the Institute of Locomotive Engineers in 1914-5. He was Hon. Sec. of the GER Railway Ambulance Service and took a keen interest in its training. He retired at the Grouping in 1923 and became a Justice of the Peace initially in West Ham, later Bexhill-on-sea, where he died suddenly while on the golf course on 14 March 1927, aged 65.

Samuel Johnson

Samuel Waite Johnson was born in Bramley, Leeds, on 14 October 1831, the son of James Johnson, an engineer who later worked for the Great Northern Railway. Samuel was educated at Leeds Grammar School and became an engineering pupil of James Fenton at the Leeds firm of E.B. Wilson. His first appointment was as Assistant District Locomotive Superintendent of the Southern Area of the GNR, then Works Manager at Peterborough and in 1859 he became Acting Locomotive Superintendent of the Manchester, Sheffield and Lincolnshire Railway, one of the companies that later formed the Great Central.

In 1864 he moved to Cowlairs, Glasgow, as Locomotive Superintendent of the Edinburgh & Glasgow Railway, which was amalgamated with the North British Railway in 1865. For his service of training pupils for the Egyptian railways at Cowlairs, the Viceroy of Egypt made him a Commander of the Order of Medjidie and an Officer of the Order of Osmanieh. In 1866 he went south to Stratford as Locomotive Superintendent of the Great Eastern Railway after the resignation for health reasons of Robert Sinclair.

He was chosen from twenty-six applicants by the Directors of the Midland Railway to be appointed as Locomotive Superintendent after the death of Matthew Kirtley, at an initial salary of £2,000 a year (compared with £750 on the GER) raised to £2,500 a couple of years later in 1875. He remained in charge of locomotive matters for over thirty years. He produced his first 4-4-0 design in 1876 and numerous developments of this throughout his period of office.

Samuel Johnson retired at the end of 1903, aged 72, and lived at Nottingham out of the limelight where he was a Justice of the Peace and was very involved with St Peter's church there. He became a Member of the Institute of Mechanical Engineers in 1861, the Institute of Civil Engineers in 1867, a Council Member of the IMechE in 1884, Vice President in 1895 and President in 1898. His son James followed in his father's footsteps, becoming Locomotive Superintendent of the Great North of Scotland Railway in 1890. Samuel died in 1912, aged 80.

Chapter 2
THE GREAT NORTHERN 4-4-0s

There were four classes of Great Northern 4-4-0s classified by the LNER as D1–D4. Although I am going to describe them in their classification order, in fact the LNER D3s were built first from 1896, followed by the D2s from 1898 with the D1s not being introduced until 1911. The first Ivatt 4-4-0s were his '400' class of 1896, with a further twenty being constructed in 1897, another twenty in 1898 and a final ten in 1899. To make matters more complicated, Ivatt designated them as GNR class D2 although they were more generally referred to as the '400' class. In 1912, Gresley brought out a new boiler type and most of these D2s were rebuilt with it, and the rebuilds were classified by both the GNR and the LNER as class D3. The half a dozen GNR D2s that were never rebuilt were classified by the LNER as class D4. The GNR 4-4-0s of 1898, a slightly enlarged version of the '400' class, the '1321' class, were classified by the GNR as their class D1, but these were identified by the LNER as class D2. Just to complicate it further, a couple of D3s were rebuilt as D2s in the early LNER era. The Ivatt 1911 4-4-0s, the '51' class, were superheated with piston instead of slide valves and the GNR included them in their D1 category with their 1898 design, but the LNER separated out these fifteen engines and classified them as LNER D1. To summarise:

GNR 400 class D2 of 1896, rebuilt with Gresley boiler from 1912, became LNER D3, D4 if left unrebuilt.
GNR 1321 class of 1898, also known as D1 from 1900, became LNER class D2.
GNR 51 class of 1911 became GNR and LNER D1 class.

Their performance on GNR top link work was very short-lived as from around 1900 Ivatt's 4-4-2 Atlantics of both C1 and C2 classes were the mainstay of Great Northern East Coast expresses until the introduction of Gresley's A1 pacifics in 1922. Just a few runs have been traced of the later D1 class on the ECML – most surviving logs and performance accounts of GN 4-4-0s are on secondary routes in Lincolnshire, Lancashire and Scotland.

The D1, H.A. Ivatt, 1911
Design & construction
Henry Ivatt introduced his final class of 4-4-0s for the Great Northern Railway in 1911 just before his retirement and handing over the reins at Doncaster to Nigel Gresley. They were numbered 51-65 and built to Engine Order 265. They were developments of his earlier '1321' class with superheaters, piston valves and a higher pitched boiler. Their dimensions were:

Coupled wheel diameter:	6ft 8in
Bogie wheel diameter:	3ft 8in
Cylinders (2 inside):	18½ x 26in
Stephenson valve gear with piston valves	
Boiler pressure:	170lbs psi
Heating surface:	1,129sq ft (of which superheater 192sq ft)
Grate area:	19sq ft
Axleload:	18 tons
Weight (Engine):	53 tons 6 cwt
(Tender):	43 tons 2 cwt
(Total):	96 tons 8 cwt

Water capacity:	3,500 gallons
Coal capacity:	6½ tons
Tractive effort:	16,074lbs

The LNER added 3000 to the numbers of former GNR locomotives in 1923, so the D1s became 3051–3065. In the LNER 1946 renumbering scheme the fifteen engines were allocated 2202–2216, although only ten survived long enough the receive these numbers.

The engines were built with Schmidt superheaters, although seven were replaced with Robinson superheaters between 1932 and 1934. They were recognisable from the earlier GNR 4-4-0s by having extended smokeboxes to incorporate the superheater elements. Most had Ramsbottom type safety valves, but 60 and 61 were equipped with Ross type pop safety valves. The cylinders' diameters were an inch larger than the earlier GN 4-4-0s and had 8in piston valves. The locomotives were fitted with Wakefield mechanical lubricators placed on the right-hand side of the running plate.

The locomotives had standard vacuum brakes, but in 1925 the class was sent to Scotland and equipped with Westinghouse brakes and had their boiler mountings lowered to fit the North British loading gauge, reducing the maximum height from a fraction under 13ft 4in to 12ft 11in. The placement of the Westinghouse brake pumps high on the right-hand side of the smokebox interfered with the driver's vision, so this was transferred to the left-hand side and later lowered to ease oiling. Seven locomotives were transferred from the Scottish Division to the Great Eastern Section between 1930 and 1932 and retained their

D1 57 as built in 1911 on a local train at Basford, c1920.
(G. Gillford/R.K. Blencowe/John Scott-Morgan Collections)

Westinghouse air brakes until 1937 when they were converted back to vacuum brakes, though two (3053 and 3059) retained their Westinghouse brakes as well for a couple of years. Six of the eight engines remaining in Scotland were altered from Westinghouse to steam brakes during the war years.

The GNR 3,500 gallon tenders were replaced by alternative types during Works visits. Two (3060 and 3065) gained 3,670 gallon tenders from a couple of Ivatt single-wheelers. 3052/3 and 3061-4 had 3000 gallon tenders that were built in 1905. There were subsequent changes between members of the class.

Their GNR livery was replaced by LNER lined green, but they were repainted black by Cowlairs Works when transferred to Scotland. Seven of the class survived to be taken into British Railways stock in 1948, but only three had the Eastern Region 60,000 number added, 62203/08/15. Only 62208 wore a smokebox door numberplate. The first D1 withdrawal was of 3051 in February 1946, four more being taken out of service that year. Four were still extant in 1950 – the three that had been renumbered – but the final member to be withdrawn was 2209, still un-renumbered, in November of that year.

D1 3059 in the early 1930s, one of the D1s to return from Scotland to the GE Division in 1931. It is seen with Ross pop safety valves, Wakefield mechanical lubricator and reduced cabside cut-out. (W.A. Camwell/MLS Collection)

D1 3058 at Carlisle Canal in September 1934. It was fitted with Westinghouse air brakes on transfer to Scotland and the pumps are here in their final position low on the left-hand side of the smokebox. It retains its Ramsbottom safety valves. It was renumbered 2209 in 1946 and was the last D1 to be withdrawn in November 1950. (MLS Collection)

D1 3064 at its home depot, Haymarket, 10 April 1939. It retains the Ramsbottom safety valves and larger cabside cut-outs. 3064 remained in Scotland during the war shedded at Perth. As 62215, it was withdrawn from Stirling in February 1950. (MLS Collection)

The last survivor, 2209, in store at Stirling in August 1950. It was withdrawn in November. It has pop safety valves and short dome cover, and steam brake replacing the Westinghouse system. (O. Carter/MLS Collection)

Operation

The initial allocation of the D1s was as follows:

King's Cross:	59–62
Doncaster:	63–65
Leeds:	51–58

When they were constructed, the East Coast main line expresses were primarily in the hands of the large boilered Ivatt C1 Atlantics. The D1s were employed mainly on lighter trains or secondary semi-fast services. However, the Leeds and Doncaster allocated engines put in much work on heavy expresses between Doncaster and Leeds, which could load up to fifteen bogie vehicles. They also had regular turns on expresses especially between Grantham and York, particularly on the heavy 13-coach 2.20pm King's Cross–Edinburgh. A couple of logs were found in the archives of the Railway Performance Society, both with No.58 shortly after its construction in 1911. There was a reallocation of members of the class during the First World War, with a concentration of the D1s at Peterborough and Grantham, the latter receiving the whole King's Cross allocation and Peterborough gaining 56–58 from Leeds. The undated logs below are likely to relate to the period shortly after their construction, as just Leeds based Nos. 56 and 58 were identified in the Railway Performance Society archives on this work.

Two other runs around this time were recorded, another with 58 also completed in 96 minutes 20 seconds and with 56 in exactly the scheduled 97 minutes. The King's Cross D1s had previously been used for stopping and semi-fast trains to Cambridge and Peterborough and were used as pilot locomotives to the Atlantics on the heaviest East Coast expresses, the Leeds and Doncaster engines often similarly used north of Grantham. Both these depots used their fleet on the expresses between Doncaster and Leeds as well as secondary services in South Yorkshire.

After the war, the D1s lost their hardest work on the Leeds route to the powerful Gresley K3 2-6-0s and some of the Great Central 4-6-0 classes and King's Cross regained 56, 60 and 62 from Grantham to resume work on the Peterborough and Cambridge stopping passenger

Grantham–York

2.20pm King's Cross–Edinburgh

Miles	Location	58 13 chs, 388/410 tons			58 13 chs, 395/415 tons			Gradients
		Times	Speeds		Times	Speeds		
0	Grantham	00.00		T	00.00		T	
4.2	Barkston	07.50			08.10			1/200 F
	Hougham	-	72½		-	69		1/300 F
9.9	Claypole	12.50			13.20			
14.6	Newark	16.55			17.35			L
20.9	Carlton	22.35	66		23.35	60		L
26.4	Tuxford	28.45			30.15			1/200 R
	Markham	-	41½		-	37		1/200 R
33.1	Retford	36.35			38.25	64½		1/200 F, 1/178 F
38.4	Ranskill	42.00	58/65		44.00	58		L
42.2	Bawtry	45.30	60		47.50	60/42		1/198 R
50.5	Doncaster	56.00	sigs	2 E	57.10	50	¾ E	L
57.5	Moss	-	sigs		65.00	58		L
64.3	Templehurst	-	sigs		72.20	56		L
68.9	Selby	-	sigs		77.55			L
75.6	Escrick	-	sigs		87.15			
80.7	York	(94.00 net)			96.20		¾ E	

D1 No.59, one of the four based at King's Cross, on a 13-coach down suburban train to Hitchin and either Peterborough or Cambridge, passing Hadley Wood, c1912. (H. Gordon Tidey/ MLS Collection)

and parcels trains. In 1923 after the Grouping the allocation was:

Cambridge:	56
New England (Peterborough):	63–65
Grantham:	57–62
Leeds:	51–55

With the influx of Gresley's Pacifics taking over much of the express work of the C1 Atlantics, these in turn made the D1s redundant on the former Great Northern metals and a decision was made in 1925 to reallocate the whole fleet of fifteen locomotives to Scotland to replace some of the ageing North British and Great North of Scotland 4-4-0s. The initial Scottish allocations were:

Haymarket:	3051, 3060, 3062 – 65
Dunfermline:	3052
Ladybank:	3053, 3055
Hawick:	3054
Carlisle Canal:	3056, 3058
St Margaret's:	3057
Eastfield:	3059, 3061

They were used primarily on local stopping trains, though they were also used to pilot heavy expresses from Edinburgh towards both Dundee and Newcastle. The locomotives were not popular with Scottish crews and by the 1930s were only used on menial duties, local passenger, goods and engineering train duties. They gained a reputation for rough riding, sparse and uncomfortable cabs compared with the NB and GNoS engines they replaced and they were often in trouble for steam. Being unpopular and only used on light duties, they were not given priority for maintenance which exacerbated their unreliability. I have discovered one log of a D1 in Scotland (no date given) with a Haymarket example on the Glasgow–Edinburgh Waverley route.

Glasgow Queen St–Edinburgh Waverley

3065

195/205 tons

Miles	Location	Times	Speed		Gradients
0	Glasgow Q St	00.00		T	1/41 R
1.5	Cowlairs	04.47			(banked by an N B 0-6-2T)
6.2	Lenzie	11.10			L
11.4	Croy	18.20	pws		1/900 R
15.5	Castlecary	22.52	58		L
21.8	Falkirk	29.49	55/ pws		L
25	Polmont	33.32			
29.7	Linlithgow	38.05	62		1/882 F
35.3	Winchburgh Jn	44.23	52½		1/882 R
39.1	Ratho	48.32	60/65		1/960 F
43.8	Saughton Jn	53.29	61½		1/960 F
46.1	Haymarket	56.07			
47.3	Edinburgh	60.07 (59 net)		T	

By 1930, it seems that the LNER Scottish Division had had enough of them and 3053, 3056, 3059 and 3062 were despatched back south of the Border and 3052, 3055, and 3060 followed in 1932, leaving just 3054 at Haymarket, 3062 at Perth and the remainder at Carlisle Canal at the west end of the Waverley route. A couple – 3051 and 3054 – were then loaned to the NE Section for trial working in the Newcastle–Middlesbrough area and the seven sent earlier were stored at Doncaster before being transferred to the former Great Eastern area, based at Ipswich, Norwich and Peterborough East. The latter's allocation was short-lived, their engines being relocated to New England and Boston. By 1936, all the Southern area engines were based at Norwich on local branch train workings where their Westinghouse braking system was

One of the D1s sent to Scotland, possibly St Margaret's 3057, crossing the Forth Bridge with a Dundee–Edinburgh, stopping train, c1928. (Rail Archive Stephenson/John Scott-Morgan Collections)

One of the Carlisle Canal engines, 3058, at Silloth with a stopping service of thirteen 6-wheelers, c1930. Note that the Westinghouse pump has been relocated on the left-hand side of the smokebox but not yet lowered to its final position. (Locomotive & General/MLS Collection)

compatible with the former GE locomotives. The 1939 allocation was as follows:

Carlisle Canal:	3058, 3061
Hawick:	3057, 3063
Haymarket:	3051, 3054, 3064, 3065
Norwich:	3052, 3053, 3055, 3056, 3059, 3060, 3062

During the Second World War, five D1s spent a year at Cambridge before returning to their Norwich base and spending most of their time on the former Midland & Great Northern Railway where they were better received than they had been in Scotland. At the end of the war, the allocation was as follows:

Hawick:	3051, 3057, 3065
Haymarket:	3058, 3061, 3063
Perth:	3054, 3064
Norwich:	3056, 3059, 3060
Yarmouth Beach:	3052, 3053, 3055, 3062

3051, 3053, 3055, 3060 and 3062 were withdrawn in 1946 and 3059, 3061 and 3065 in 1947. The remainder were renumbered with 2203 (3052) at Norwich, 2205 (3054) at Dunfermline, 2207 (3056) at Yarmouth Beach, 2208 (3057) at Hawick, 62209 (3058) at Stirling, 2214 (3063) at Haymarket and 2215 (3064) at Perth. The two GE Section engines finished their careers after nationalisation at Norwich (2207) and Colwick (62203). They did little work and spent time in store before the class became extinct in 1950.

The D2, H.A. Ivatt, 1898

A slightly enlarged version of Ivatt's first 4-4-0 for the Great Northern Railway was the '1321' class which the GNR classified as class D1 in 1900. 1321–1335 were constructed at Doncaster in 1898, another ten (1336–1340 and 1361–1365) in 1899, 1366–1385 in the next Lot built between 1900 and 1901, ten more (1386–1395) in 1903, another five (1396–1399 and 1180) in 1907 and the final ten of the seventy strong class in 1909, numbered 41–50. The boiler was 3in larger in diameter and the grate area was increased by 17 per cent. Their main dimensions were:

Coupled wheel diameter:	6ft 8in
Bogie wheel diameter:	3ft 8in
Cylinders (2 inside):	17½ x 26in
Stephenson valve gear with slide valves	
Boiler pressure:	175lbs psi
Heating surface:	1,253sq ft (1,129sq ft when later superheated)
Grate area:	19sq ft (20.8sq ft for the 1898 batch only)
Axleload:	17 tons
Weight (Engine):	47 tons 10 cwt
(Tender):	40 tons 18 cwt
(Total):	88 tons 8 cwt
Water capacity:	3,670 gallons
Coal capacity:	5 tons
Tractive effort:	14,805lbs

1321–1325 were perceived as having inadequate smokebox volume causing initial steaming problems and this was corrected from 1326 onwards, with a repositioning of blastpipe and chimney. 1331 was equipped experimentally with Marshall's valve gear for six months in 1902 before the experiment was abandoned. There were minor detailed variations in appearance in the engines from 1903 onwards, mainly running plate and handrail design issues.

There were various experiments carried out at Ivatt's instruction with spark arrestors and superheaters, with Gresley deciding to fit one member of the class (1381) with a Robinson superheater in 1914, at the same time lowering the boiler pressure to 170lbs psi. The superheated engine then received an extended smokebox similar to the 1911 built D1s. The LNER renumbered the locomotives, as with all the ex GN fleet, with an additional 3000, and reclassified them as class D2 to differentiate them from the 1911 built D1 locomotives.

Left: **GNR 1370**, built in November 1900, seen as built in the first decade of the twentieth century, but with smaller 3,060 gallon tender.
(F. Moore/MLS Collection)

Below: **One of** the first 1898 locomotives now reclassified as the LNER D2 class, in the initial LNER lined green livery, taken at King's Cross in 1923 before the addition of 3000 to GN engine numbers.
(F. Moore/MLS Collection)

4396, built in 1907, with slightly enlarged extended smokebox, newly out-shopped from Doncaster Works, in its LNER livery of 1924. (MLS Collection)

The original prototype, 1321, now 4321, as reboilered and seen in the company of a Gresley O2 2-8-0, at Doncaster, c1930. Note that the cabside cut-out gap has been reduced. (Photomatic/MLS Collection)

Twenty-three of these locomotives were subsequently rebuilt with superheaters like 1381 in the 1930s, the last (3050) being so equipped in 1937. Two of the GNR '400' class (then classified as D3s) were rebuilt as D2s, 1305 (4305) in 1923 and 4320 in 1926, assumed to be because of the availability of D2 type frames when those two locomotives needed major repairs to theirs. There seems to have been a crisis in the availability of new boilers for the D2s in 1928 with the result that many lay for several months in Doncaster Works. Some superheated boilers reverted to the saturated type and ten were fitted to D2s to expedite their return to traffic, three more were fitted with superheated boilers and ten new saturated boilers were constructed and fitted to D2s in 1929. A number of locomotives had the height of their chimneys reduced to enable them to operate over GE routes which had restricted height loading gauge. Ross pop safety valves replaced the Ramsbottom version on new boilers constructed after 1926. The depth of the cabside cut-out was reduced by 9in between 1929 and 1933.

Various GN type tenders were used to replace the original 3,670 gallon ones during Works visits, with types holding 3,000 and 3,170 gallons, as presumably their long distance work reduced the need for the larger capacity as Gresley's new designs held sway. Those engines finishing up on the Midland & Great Northern section after 1937 received tablet-exchange apparatus.

The locomotives received the standard GN passenger lined green livery until the Grouping, with the change to the LNER green with number on the tender from 1923 causing an occasional mismatch between engine and tender identity. From 1928, their livery changed to black with red lining, the numbers being removed from the tender to the cab shortly afterwards. During the Second World War the livery was plain black with just the initials NE on the tender.

Thirty-one of the seventy D2s became part of BR's stock in 1948. They had been renumbered in the LNER 1946 scheme between 2152 and 2201 and only one, 62172, lasted long enough to receive its nationalised number and BRITISH RAILWAYS on the tender. The first withdrawal took place (4334) in 1936 and nineteen were withdrawn before the start of the Second World War. The increased wartime traffic requirements kept them all in service until 1946, when withdrawals recommenced in earnest. Five survived to 1950 (2154, 2161, 2173, 2180) and 62172 was withdrawn in June 1951.

Operation

Initially these Ivatt 4-4-0s were scattered throughout the Great Northern system and used mainly on secondary passenger services, augmenting the numbers of the earlier '400' class, especially in Nottinghamshire and Lincolnshire. In the early days, some were used on East Coast expresses from King's Cross often with assistance from a 2-4-0 or Stirling single-wheeler. Charles Rous-Marten wrote his first

3041 and 3042 were based at Hatfield for several years working Dunstable–King's Cross trains but they were replaced and this shot of 3042, a 1909 built engine, is seen a couple of years later somewhere on the GE section, with shorter chimney and reduced cabside cut-out.
(W.A. Camwell/MLS Collection)

D2 2162 (formerly 1335/4335 of 1898) at Colwick shed, 27 July 1947, just withdrawn that month. (J.D. Darby/MLS Collection)

The last survivor and the only D2 to receiver its BR number, 62172 (formerly 1369/4369 of 1900) at Doncaster in the company of a former Great Central A5 4-6-2T, c1950. It was withdrawn in July 1951. (J.D. Darby/MLS Collection)

article *British Locomotive Practice and Performance* in the *Railway Magazine* of September 1901, and the early articles were much obsessed with the possible replacement of the 'singles', especially those of the Great Western, Midland and Great Northern for which Rous-Marten seemed to have great affection, by the increasing use of four-coupled engines which he was having to admit were needed to cope with increasing train weights.

As early as December 1901 he described a couple of runs by GN D1s of the '1321' series on the 9.45am King's Cross–Leeds with 200-ton trains, scheduled to pass Peterborough in 80 minutes for the 76.2 miles and 115 minutes for the 105 miles to Grantham. 1377 built in January 1901 ran smartly to Potters

Bar, 12.7 miles, in 17 minutes 34 seconds and had run the 72.5 miles to a signal stop at Yaxley in 73 minutes 10 seconds, passing Peterborough in 81 minutes 16 seconds and arriving at Grantham in a net 109 minutes. 1338, built in 1899, passed Potters Bar in 17 minutes 9 seconds with 50mph at the top of the 1 in 200, climbed to Abbots Ripton (1 in 200) at 55mph, passed Peterborough in just ten seconds over 80 minutes and after a signal check at Tallington, accelerated to 52mph at the top of the 1 in 178 to Stoke Tunnel. Net time to Grantham was 110½ minutes. 1383 with a 13-coach 250 ton load on the 1.30pm King's Cross–Leeds took nearly 20 minutes to clear Potters Bar at 43mph but then recovered and got to Peterborough in exactly 80 minutes, arriving three minutes early, an average of 57.2mph, with a minimum of 50mph at Abbots Ripton. Downhill speeds were not excessive, not much over 70mph. In 1903, 1365 took over at Doncaster from the prototype Ivatt Atlantic, 990, running the 175-ton train at 60mph on the 1 in 200 before South Elmsall and 56mph on the 1 in 150 to Nostell, arriving at Wakefield (19.9 miles) in 21 minutes 16 seconds.

In the previous section of this book there was a log table demonstrating the 1911 D1's use on the 2.20pm King's Cross–Edinburgh between Grantham and York. The 1909 series of the D2s also participated in the working of this train and a log of a run with No.41 which kept exact time is shown below.

Grantham–York
2.20pm King's Cross–Edinburgh
41
13 chs, 402/425 tons

Miles	Location	Times	Speeds		Gradients
0	Grantham	00.00		T	
4.2	Barkston	08.00			1/200 F
	Hougham	-	69		1/300 F
9.9	Claypole	13.10			
14.6	Newark	17.25			L
20.9	Carlton	23.35	65		L
26.4	Tuxford	30.10			1/200 R
	Markham	-	32		1/200 R
33.1	Retford	38.30	60		1/200 F, 1/178 F
38.4	Ranskill	44.00	58/65		L
42.2	Bawtry	47.45	60/45		1/198 R
50.5	Doncaster	57.15	* slow	¾ E	L
57.5	Moss	65.35	50		L
64.3	Templehurst	73.05	55		L
68.9	Selby	78.35	*		L
75.6	Escrick	87.25			
80.7	York	96.50		¼ E	

41 is recorded on a second occasion making the exact same overall time. No speeds were shown in the log, but I have estimated key ones comparing the times with the second run by 1911 D1 No.58. The series 41–50 seem to have replaced the small boilered C2 Atlantics on light express and piloting duties during the First World War years. Like the 1911 D1s, they worked express portions between Doncaster and Leeds and across the GC route to Manchester with GN services. The proliferation of fast braked goods trains saw the use of these engines increasingly until Gresley 2-6-0s were available in significant numbers. With the heavy loads in the period immediately after the First World War, before the increase in frequency and reduction in train weight, the D2s were frequently employed assisting the Ivatt C1 Atlantics. The King's Cross D2s worked suburban trains from Hitchin and Hatfield – 1385 is recorded as taking 22 minutes for the fifteen mainly downhill miles to the Finsbury Park stop on the 5.48pm Hatfield with just six coaches.

The allocation in 1923 as the LNER took over the pre-grouping rolling stock was:

King's Cross:	13
Hitchin:	3
Grantham:	5
Colwick:	12
Boston:	10
Lincoln:	6
Louth:	1
Retford:	3
Doncaster:	6
West Riding:	5
York:	6

GNR D1 1381, built in 1901, on a Down East Coast express in the Potter's Bar–Hatfield area, c1903. (F. Moore/MLS Collection)

GNR D1 No.50, on a down East Coast express at Harringay, c1910. (Loco Publishing Co./MLS Collection)

D2 1386, built in January 1903, approaching Doncaster from the North with a heavy class 'H' unfitted freight, c1920. (MLS Collection)

D2 1336, built in 1899, assisting C1 Atlantic 1406 at Belle Isle on a Down heavy East Coast express, climbing towards the Copenhagen Tunnel, 1923. (Real Photographs/MLS Collection)

3044 with a Nottingham Victoria–Derby Friargate stopping train near Breadsall, c1925. (MLS Collection)

King's Cross D2s acted as standby engines there ready, with a C1 Atlantic, to cover failures or provide assistant power. By 1930 there were sufficient Pacifics to render such standby engines unnecessary. The West Riding and Doncaster engines worked local passenger services in the area and Doncaster engines also worked to Leeds, Hull, Lincoln and Peterborough. Engines in this region also double-headed East Coast expresses north from Grantham until the Gresley Pacifics were more numerous. The main area of their work in the early LNER days was in Lincolnshire and on slow trains to the coastal towns of Skegness, Scarborough, Cleethorpes and Bridlington. Five (4361/78/81/92/99) were transferred to Stockport and Trafford Park in 1932-33 to operate over the Cheshire Lines. The allocation on 1 January 1934 was:

King's Cross:	5
Hatfield:	2 (3041 & 3042)
Hitchin:	8
New England:	3
Grantham:	7
Colwick:	13
Boston:	8
Lincoln:	7
Louth:	3
Retford:	5
Doncaster:	1
York:	5 (4180, 4386/87/96/98 – three went to Hull Botanic Gardens in 1937)
Stockport:	2
Trafford Park:	3

Just as the Scottish D1s were unpopular, so the NE York based D2s were equally disliked when compared with the D20s they replaced. D2 4381 was timed by Albert Mellor on the 3.30pm Manchester Central–Liverpool on 23 February 1935 with seven coaches (168/183 tons) and comfortably beat the schedule, the 15½ miles to Warrington, schedule 18 minutes run in 16¾, the 4.2 miles from Glazebrook to Padgate with stretches of 1in 240 falling mixed with level patches tempting the D2 to a maximum of approximately 68mph. The 21 miles on to Liverpool Exchange, including a one minute stop at Farnworth, was concluded in 24 minutes 5 seconds start to stop, (net 23½) against a schedule of 26 minutes. The rush down the 1 in 185 from the Farnworth start meant that the 1.7 miles in the dip (1/185 down–1/185 up) was covered in exactly 1½ minutes and the descent from Hunts Cross to Mersey Road produced another high '60'.

Stopping trains noted in the Nottingham/Lincoln area between February and April 1936 included 4392 between Spalding and Boston and 4331 on a Boston–Lincoln stopper on 25 February 1936, 3044 on a Lincoln–Grantham train and 4329 from Newark to Nottingham on 27 February, 4362 on a Nottingham–Derby Friargate train on 21 April and 3043 on a return trip from Nottingham to Grantham and 4399 on to Boston on St George's Day. In 1935, several were in store, but in 1937, with the take over by the LNER of the M&GN routes in North Norfolk, twelve D2s and ten D3s were drafted in, replacing many of the ageing M&GN former Midland power. Drivers seemed to get better work out of them here. 4374 was timed by the Rev R.S. Haines on the six coach 10.26am Melton Constable to Peterborough on 18 August 1937 taking 12 minutes 53 seconds for the nine miles from South Lynn to Sutton Bridge – a slight loss – and 10¾ minutes on to Wisbech (schedule 11½). Thorney, 11.7 miles from Wisbech, was passed in 16¾ minutes (just 11 seconds late) but signal checks outside Peterborough made the train a minute late on arrival (30 minutes 6 seconds for the 20.1 miles). 3042 was noted on the 11am King's Lynn–North Walsham calling at Corpandy, Aylsham and Fakenham. 4373 on a Norwich–Leicester train recorded by Gerald Aston in 1938 ran the 9½ miles from Melton to Fakenham in 13¾ minutes (schedule 14) with a maximum speed of 62mph. Fakenham–South Lynn (22.1 miles

4375 at Newark on a Doncaster–Grantham stopping train, c1938.
(Photomatic/MLS Collection)

was covered in 32 minutes 20 seconds with a maximum of 56mph before signal checks approaching destination (schedule 32 minutes). At South Lynn, the train was taken over by Midland 2P 4-4-0, No.542. Withdrawals commenced in 1936 and by 1939 twenty had been scrapped. The allocation at the start of the Second World War in September 1939 was:

Hitchin:	3
New England:	4
Grantham:	6
Colwick:	8
Boston:	10
Lincoln:	3
Louth:	2
Retford:	1
Doncaster:	1
Frodingham:	1
Botanic Gardens:	3
Bury St Edmund's:	1
South Lynn:	4
King's Lynn:	1
Melton Constable:	2
Yarmouth Beach:	2

Because of the increase in wartime traffic, withdrawal of the D2s ceased until 1946, working local passenger services thus releasing larger engines for heavier work. Despite their unsuitability, some found work on essential coal and other freight services. With the enormous loads on the East Coast during the war, some found duty as assistant engines to Gresley Pacifics. At the end of the war, in poor repair and unloved, some twenty-one were condemned in the two years before nationalisation.

The following was the allocation as the remaining D2s were taken into British Railways ownership:

Hitchin:	2
New England:	1
Grantham:	3
Colwick:	13
Boston:	4
Melton Constable:	6
Yarmouth Beach:	2

Withdrawals then took place quickly. Two D3s had been rebuilt as D2s in the 1920s and were both withdrawn in 1949 in addition to another eleven. Just six made it to 1950, only 62172 acquiring its BR number. 2154 and 2181, allocated BR numbers but not carried were withdrawn in November 1950, 62172 in June 1951, making the class extinct.

D2 2161 (built as GN D1 1333 in 1898) at Nottingham Victoria alongside a GC C4 Atlantic, c1946. (John Scott-Morgan Collection)

The last surviving D2, 62172, on a Grindley–Derby milk train, 25 June 1949. (W.A. Camwell/MLS Collection)

2190 (former 4395) after withdrawal at Doncaster and awaiting scrapping, 17 May 1947. (MLS Collection)

The D3 & D4, H.A. Ivatt, 1896

Henry Ivatt became the GNR's Locomotive Engineer in March 1896, and he had designed and constructed the company's first 4-4-0 by December of that year. It was numbered 400 and was similar in many ways to the E1 2-4-0s (the final series, 1061-1070, was brought out simultaneously). Its dimensions were:

Cylinders (2 inside):	17½ x 26in
Coupled wheel diameter:	6ft 8in
Bogie wheel diameter:	3ft 8in

Stephenson valve gear with slide valves

Boiler pressure:	175lbs psi
Heating surface:	1,123sq ft
Grate area:	17.8sq ft (reduced to 16.25sq ft from 1907)
Axleload:	14 tons 9 cwt
Weight (engine):	44 tons 7 cwt
(tender):	38 tons 10 cwt
(total):	82 tons 17 cwt
Water capacity:	3,170 gallons
Coal capacity:	6 tons
Tractive effort:	14,805lbs

A further fifty engines were built to this design, 1071–1080 and 1301–1310, in 1897, 1311–1320 and 1341–1350 in 1898 and 1351–1360 in 1899, the latter thirty at the same time as the larger '1321' class indicated by the number gaps in the 13XX series filled by the slightly larger engines. In 1900, the 1896 designed engines became identified as the D2 class although most referred to them still as the '400s'. On the first batches the running plate was flat, but on 1320 and 1351–1360 the running plate was raised over the coupled wheels in the same style as that introduced on the Atlantics. The last ten were also provided with larger 3,670 gallon tenders – outwardly similar but with a 500 gallon tank between the frames. Four retained this to their withdrawal. Subsequently, tenders changed between classes at Works visits, including the use of old Stirling tenders and the 3,000 gallon tenders from other Ivatt 4-4-0 classes.

A year after his arrival at Doncaster, Nigel Gresley designed an enlarged boiler with a 4ft 8in diameter barrel compared with the previous 4ft 5in, though still with a short smokebox, and fitted it to 1359 in 1912. A shorter chimney was fitted. The other variations from the 1896 design were:

Heating area:	1,235sq ft
Weight (engine):	45 tons 14 cwt
(total):	84 tons 4 cwt

GNR '400' class 1077 built in 1897 and classified D2 in 1900, seen here in its original condition, c1910. It remained in this condition until the grouping and was then identified as one of the six D4s, the rest being rebuilt with larger boilers by Gresley between 1913 and 1921. It was then rebuilt as a D3 in June 1923 and was withdrawn in 1937.
(E. Pouteau/MLS Collection)

Above left: **GNR 1348,** as built in December 1898 with large 3,670 gallon tender. It was renumbered 4348 in 1924, 2171 in 1946 and was withdrawn in 1947. (MLS Collection)

Above right: **1354 built** in November 1899, as a GN D2 in original condition, at Nottingham Victoria, 17 May 1912. (John Scott-Morgan Collection)

Left: **1358, one** of the last batch of GNR D2s built in 1899, still in unrebuilt condition, c1920. It has the raised footplating over the coupled wheels, a tall dome cover and a slightly extended smokebox, but remained in this condition and classified by the LNER as a D4 until rebuilt with the Gresley enlarged boiler in 1928 as a D3, the last to be converted. (Real Photographs/MLS Collection)

Most of the other D2s were rebuilt with the Gresley boiler between 1913 and 1921, but retaining the original longer chimney and were identified by the GNR and subsequently by the LNER as class D3. Most later gained shorter chimneys. 1316, rebuilt in February 1914, had separate splashers instead of continuous ones over the coupled wheels but this variation remained unique for this class. They were renumbered at the Grouping to 3400, 4071–4080, 4301–4320 and 4341–4360. The GNR two-tone green livery was replaced at the Grouping with the lined LNER green and numerals on the tender until 1928 when they were painted black with red lining. The last to retain the LNER green livery was 4346, which remained green until 1932.

1080, built as a '400' class in 1897, classified as a D2 in 1900 and rebuilt as a D3 in 1918. It is seen here at King's Cross in the new LNER livery but without renumbering, c1924. It retains the tall chimney, deep cab cut-out and Ramsbottom safety valves. (MLS Collection)

The prototype
1896 Ivatt '400' 4-4-0, after rebuilding as a D3 in 1920 and repainting and renumbering 3400 in the lined green LNER livery, ex-works at Doncaster in May 1925. It was withdrawn in 1947. (MLS Collection)

The six still unrebuilt in 1923 – 4077, 4079, 4313, 4356, 4358 and 4360 – remained unrebuilt in their GNR D2 guise and were given the new classification D4 in 1923. They were all rebuilt to the D3 class by the LNER in the 1920s, the last conversion being of 4358 in 1928. Like the other Ivatt 4-4-0s, their Ramsbottom safety valves were replaced gradually with the Ross pop type. Consideration was given to equipping them with Robinson superheaters and orders were made, though subsequently cancelled. Two D3s, 4305 in 1923 and 4320 in 1926, were rebuilt as D2s – an extensive rebuilding involving frames as well as boiler. It is understood that both engines had required substantial repairs to their frames and the opportunity was taken to finish them as the slightly larger engine.

Complaints of draughty cabs were made in the late 1920s, resulting in a reduced cut-out in the cabside being approved and this alteration was made to all the former GNR 4-4-0s as they went through Works for general overhaul. Four engines, 4075, 4077, 4349 and 4354, were given cabs with no cut-outs in 1935 for use on the Stainmore line over the Pennines, but two were with drawn in 1937 and 4075 and 4349 returned south. 4075's cab was transformed in 1944 when it was allocated the duty of hauling officers' inspection saloons. It was provided with a cab with two windows on each side, initially renumbered '1', but swiftly altered to '2000' and, despite it still being wartime, was painted in full LNER lined green livery, a livery it continued to carry into the days of British Railways. Tablet exchange apparatus was fitted in 1937 for those locomotives allocated to the former M&GN section.

Fourteen D3s were withdrawn before the Second World War and just one, 4304, during the war in 1942 – war damage? – and nineteen were still extant on 1 January 1948 when they were taken into British Railways stock. However, only three received their BR numbers – one, 62131, the engine with separate splashers – and only the 'special' 62000 received a smokebox door numberplate. One, 2140, received the 'E' prefix and the legend 'BRITISH RAILWAYS' on the tender.

1316 rebuilt in 1914 as a D3 uniquely with separate splashers over the coupled wheels. It was renumbered 4316 after the Grouping and is seen at New Holland on 17 July 1938 in black livery. (MLS Collection)

1899 built 4351, with raised footplating, short chimney and smaller cab cut-out in LNER black livery, at Spalding with the 8.30am to Sutton Bridge, 1938. (MLS Collection)

1897 built 1075, rebuilt as a D3 in 1916, renumbered 4075 in 1924 and new side-window cab provided in 1944 when it was painted in green LNER livery and renumbered 2000 for working officer saloon specials. (MLS Collection)

2000 again, but in shabby condition awaiting Works overhaul at Doncaster, 21 September 1947. (C.H. Owen/MLS Collection)

Now BR 62000, but still repainted in LNER apple green and still used for officer saloon duties, it is on Grantham shed with A3 60053 *Sansovino* in the background, 4 June 1950. (H.C. Casserley/MLS Collection)

2126, formerly 1309 built in 1897, rebuilt as a D3 in 1918, renumbered 4309 in 1926 and 2126 in 1946, rests with another D3 at Doncaster awaiting its fate a month after its withdrawal, 19 September 1948. (J.D. Darby/MLS Collection)

Operation

These locomotives were primarily built for secondary passenger work, as the '1321' class being built simultaneously were preferred for the faster trains and they also in any case were soon superseded by the C1 and C2 Atlantics. However, before either the '1321s' or the Atlantics were in traffic, the pioneer, 400, did record a scheduled 98 minute run over the Grantham–York 80.7 mile section, though with only 235 tons. The 1897 built 1312 was timed in 1902 on the 2pm Leeds–King's Cross from Grantham by Rous-Marten. With a gross load of 180 tons, it passed Stoke in 8 minutes 53 seconds for the 5½ mile climb from the Grantham start, and with nothing over 70 down to Essendine, still passed Peterborough at slow speed in 31½ minutes and reached London, non-stop, in 117 minutes 39 seconds, 113½ minutes net, in a severe south westerly gale. The long 1 in 200 southbound climb to Knebworth brought the speed down to 55mph at Hitchin and 47mph through Stevenage. As well as semi-fast services, they piloted other Stirling and Ivatt engines north of Grantham on both Leeds and Newcastle expresses. The allocation in 1912 was as follows:

King's Cross:	6
Peterborough:	5
Grantham:	4
Colwick:	13
Retford:	7
Lincoln:	1
Doncaster:	8
Leeds:	3
Bradford:	4

Their operation varied little during the First World War and after the rebuilding of most to class D3, their 1923 LNER initial allocation was:

King's Cross:	5
Hitchin:	2
Peterborough	1
Boston:	5
Lincoln:	3
Grantham:	1
Colwick:	16
Doncaster:	4

GNR 1306, built in 1897 as a D2, on a stopping train to Nottingham Victoria at Basford, c1910, before rebuilding as a D3 in 1913. (R.K. Blencowe/John Scott-Morgan Collections)

The prototype of the '400' class at Basford on a stopping train to Nottingham Victoria, c1920. (G .Gillford/ R.K. Blencowe/John Scott-Morgan Collections)

Retford:	1
Leeds/Bradford &Ardsley:	6
York:	1 (4348)

The six unrebuilt D2s, now classified D4s, were stationed at Grantham (2), Lincoln (1) and Boston (3). The large allocation at Colwick operated stopping services to Leicester, Nottingham, Derby, Burton-on-Trent, Grantham and Newark and summer holiday services to the East Coast resorts of Skegness, Mablethorpe, Cleethorpes and Scarborough.

The King's Cross engines worked suburban services to Hatfield, Hitchin and a few stoppers to Cambridge or Peterborough and either a D2 or D3 would be one of the standby engines should a main line engine require assistance – usually if a Pacific was not available and a heavy train had an Atlantic instead.

D3 4311 assists C1 4460 with a 12-coach Down East Coast express, c1928. (MLS Collection)

The D3s in the West Riding of Yorkshire operated local services in the Leeds, Wakefield, Doncaster area and over the former GC route to Barnsley. Botanic Gardens engines worked to Beverley and Brough and slow trains to Leeds, Bridlington or York, plus empty stock working to Paragon station.

In 1930, replacements were needed for some of the 1887-built North Eastern D23s that had been working between Darlington and Penrith over Stainmore summit and 4348 was sent there on trial. Despite being considered a failure after six weeks of working, it and a further six more D3s were sent to Darlington specifically for the Stainmore route – 4075, 4077, 4347, 4349, 4350 and 4354. In addition to the Stainmore route, the Darlington D3s worked to Saltburn, Barnard Castle, Bishop Auckland and stopping trains to Newcastle. Many of these Darlington engines were sub-shedded between 1933 and 1935 at Barnard Castle (4075, 4347), Middleton-in-Teesdale (4077, 4354), Kirkby Stephen (4349) and Penrith (4077, 1935/6). The D3 working over this route appeared to end in 1935 when the closure of a section of LNER line at Penrith stopped the facility of turning the D3s (the Penrith turntable was too small and they were replaced by ex Great Eastern E4 2-4-0s which could fit).

The first withdrawals took place in 1935 and the allocation of the remaining locomotives at the end of that year was as follows:

Hitchin:	4
Peterborough:	1
Grantham:	1
Boston:	8
Louth:	4
Lincoln:	3
Retford:	2
Colwick:	11
Leeds Copley Hill:	1
Immingham:	1
Botanic Gardens:	5
York:	2
Darlington:	1
Barnard Castle:	2
Middleton-in-Teesdale:	1
Kirkby Stephen:	1
Penrith:	1

Hitchin lost its D3s to Colwick and the Stainmore route engines were transferred to Botanic Gardens and Selby. The Boston and Lincoln

D3 4313 at Hull Paragon with a local stopping train in 1934. 4313 had been one of the six D4s at the Grouping, but it was rebuilt as a D3 in 1924. (MLS Collection)

D3 4348, the engine sent in 1930 for test on the Stainmore line, seen here at Kirkby Stephen, 5 June 1935, shortly before the D3s' removal from the line.
(H.C. Casserly/MLS Collection)

D3 4075, one of the six sent to work on the Stainmore line in 1930, at Tebay NE shed, 1931. 4075 was the engine later rebuilt for officers' saloon work and renumbered 2000.
(J.A.G. Coltas/MLS Collection)

engines were occupied with local services to Skegness, Louth, Grantham and Peterborough. A major change was the transfer of the Midland & Great Northern company and stock to the LNER in October 1936. That railway still had a large number of Johnson Midland 4-4-0s of venerable age and the oldest were withdrawn and replaced by ten D3s as well as a number of D2s as indicated in the last section of this book. These ten D3s were 3400, 4302, 4306, 4310, 4315, 4319, 4345, 4352, 4355, 4356. They were allocated to New England, South Lynn and Melton Constable. 4356 (an ex-D4) was paired with former Great Central 5323 to haul the 10.26 Melton Constable–Peterborough on 18 August 1937 loaded with ten coaches and just about held schedule stopping at Fakenham, 9.8 miles in 14 minutes 10 seconds (schedule 14), Hillington, 13.7 miles in 18 minutes 53 seconds (schedule 18) and arriving at South Lynn, 7.3 miles in 12 minutes 54 seconds (Schedule 13½). The train split at South Lynn and the Peterborough portion went forward with a D2. The 3.10pm Grimsby–Peterborough in February 1936 arrived at Firsby behind a Great Central D9 and 4356 went forward with just three coaches to Lincoln arriving just 13 seconds late having stopped at eleven intermediate stations, the longest section without stopping covered in just under ten minutes. Gerald Aston timed a number of trains in the war years, including a day in September 1941 when he had runs on a number of D3-hauled local services – 4352 on the 10am Saxby – King's Lynn, where he noted a top speed of 48mph, 4304 on the 1.45pm Louth–Sutton on Sea–Willoughby (tender-first after reversal at Sutton) with just three coaches making up a couple of minutes of a ten minute late start. In May 1942 4343 was working the 4-coach 4.50pm Woodhall Junction–Horncastle which according to

Gerald left 1½ minutes early (!) and after one intermediate stop, terminated the same. They remained in the area until their withdrawal after the Second World War. At the start of the war in 1939 the allocation was:

Hitchin:	1
Peterborough:	2
Grantham:	2
Colwick:	7
Boston:	4
Louth:	1
Leeds Copley Hill:	1
Immingham:	3
South Lynn:	8
Melton Constable:	2

With the shortage of power for the increased wartime traffic, nearly all the D3s were maintained in service with just one withdrawal. The increased RAF activity in the Immingham area and American airbases in Norfolk and Lincolnshire changed the D3 distribution and at the end of the war, before mass withdrawals took place, the allocation was:

Hitchin:	3 (4073, 4309, 4359 for engineers' trains from Hitchin p-way depot)
Cambridge:	4
Grantham:	1 (2000 for Officers' saloon specials)
Lincoln:	4
Louth:	1
Langwith:	3
Staveley:	1
Immingham:	11
South Lynn:	6

D3 2135, formerly 4343, at Lincoln, 8 June 1947. 2135 was renumbered 62135 in June 1948 and was withdrawn in February 1950. (MLS Collection)

D3 2000, formerly 4075, at Grantham about to leave shed to work an officers' saloon, 8 July 1948.
(H.C. Casserly/MLS Collection)

Because of a shortage of more powerful engines, the South Lynn D3s resumed ex M&GN top link working with trains from Yarmouth to Leicester and Birmingham. Nineteen D3s entered British Railways stock on 1 January 1948, shedded at:

Hitchin:	1 (2148 – ex 4359)
Peterborough:	2
Grantham:	1 (62000)
Colwick:	6
Louth:	1 (62132)
Immingham:	2
Staveley:	1
South Lynn:	5

62148 was transferred to Colwick in May 1949 and was the last active member of the class, apart from 62000. It was working out of Derby Friargate until withdrawn in November 1950. The last observation noted of 62000 was on 23 August 1951 when it worked an officers' special north of Grantham.

Chapter 3

THE GREAT CENTRAL 4-4-0s

The Great Central built 146 4-4-0 tender locomotives between 1877 and 1922, designs by Charles Sacré, Thomas Parker, Henry Pollitt and John G. Robinson. They were all express engines, with coupled wheel diameter of 6ft 9in or 7ft, apart from the early Sacré dozen. This contrasts with the Great Northern Railway policy which basically used the 4-4-0 wheel arrangement for secondary passenger services, relying on Atlantics for the mainline long distance traffic until Pacifics became available. The Great Central did build a successful class of Atlantics (the 'Jersey Lilies', LNER C4) but these merely complemented the work of the 4-4-0s. Robinson's 4-6-0 classes intended as successors to the 4-4-0s never really fulfilled their promise and the express 4-6-0s of LNER classes B2 and B3 (later redesignated B18 and B19) remained few in number. The 4-4-0 designs of 1913 and 1919 – the D10 and D11 'Directors' – retained their role of the company's main line power until the Grouping and beyond, with Gresley perpetuating the D11s in 1924 slightly modified for use in Scotland, though for secondary roles in support of his Pacifics which were then available for the key Newcastle–Edinburgh–Dundee services (Dundee–Aberdeen remained restricted to 4-4-0s and Reid Atlantics until the route was strengthened). The descriptions of the former Great Central 4-4-0s follow, but in the LNER classification order, rather than the date order in which they were designed and constructed. The LNER D11/2s are included in this chapter as they were basically a GC design although built by the LNER to Gresley's specification. In date order the classes were:

- 1877: D12 with 6ft 3½ in coupled wheels (Charles Sacré)
- 1887: D7 with 6ft 9in coupled wheels (Thomas Parker)
- 1888: D8 with 6ft 9½ in coupled wheels (Thomas Parker)
- 1895: D5 with 7ft coupled wheels (Henry Pollitt)
- 1897: D6 with 7ft coupled wheels (Henry Pollitt)
- 1901: D9 with 6ft 9in coupled wheels (J.G. Robinson)
- 1913: D10 with 6ft 9in coupled wheels (J.G. Robinson)
- 1919: D11/1 with 6ft 9in coupled wheels (J.G. Robinson)
- 1924: D11/2 with 6ft 9in coupled wheels (Robinson/Gresley)

The D5, Henry Pollitt, 1895

In 1893, Thomas Parker was succeeded by Harry Pollitt as Locomotive Superintendent of the MS&LR. He designed a development of the Parker 4-4-0 of 1887 (the LNER D7), the latter batch of which had not been completed by Gorton Works until 1894. The drawings were completed by 1894 and the first of six new locomotives, numbered 694, did not appear until July 1895. 695–699 were completed by the end of the year. They were the first tender engines of the MS&LR to be fitted with Belpaire boilers and were designated class '11' by the GCR, D5 by the LNER at the Grouping. Their dimensions were:

Coupled wheel diameter:	7ft 0in
Bogie wheel diameter:	3ft 6in
Cylinders (2 inside):	18½ x 26in

Stephenson valve gear with slide valves

Boiler pressure:	160lbs psi
Heating surface:	1,101sq ft
Grate area:	19.59sq ft
Axleload:	17 tons
Weight (Engine):	48 tons 11 cwt
(Tender):	37 tons 6 cwt (later 43 tons)

(Total):	85 tons 17 cwt (later 91 tons 11 cwt)
Water capacity:	3,080 gallons (later 4,000 gallons)
Coal capacity:	5 tons (later 6 tons)
Tractive effort:	14,421lbs

Robinson reboilered these six locomotives between 1905 and 1913, increasing the boiler pressure to 170lbs psi, heating surface to 1,318sq ft and grate area to 20sq ft. The LNER added 5000 to the numbers of the former Great Central locomotives so the D5s became 5694–5699. After the Grouping, two examples – 5694 and 5695 – received superheated boilers, identical to those fitted to the later D6s. With 138sq ft of superheating, the revised heating surface of these two locomotives was 1,031sq ft.

Very little alteration was made to these six locomotives during their career, apart from the boiler changes. The cab roofs were extended backwards to give the crew more cover after 1912. Extended smokeboxes from around 1911 altered the appearance of them. 5694 and 5699 acquired 'flowerpot' chimneys. Ross pop safety valves replaced the Ramsbottom type on 5696 and 5698. Larger 4,000 gallon tenders were provided after the GC extension to London in 1899, though 694-697 reverted to the 3,080 gallon type as these 4-4-0s were rarely used on the London services. However, the LNER refitted them with 4,000 gallon tenders in the mid-1920s.

The locomotive livery was the GC passenger lined green, replaced by the LNER lined apple-green at the Grouping, but like many other classes, replaced by black with red lining from 1928 as an economy measure.

The first to be withdrawn was 5696 in July 1930 and all had gone by March 1933, the last survivor being 5699.

Operation

All six class '11' 4-4-0s were based initially at Gorton and between 1895 and 1897 worked Manchester–King's Cross expresses between Manchester and Grantham via the Woodhead route. The 1897 '11A' (D6) 4-4-0s displaced them on the top link work and monopolised the London work when the line to Marylebone was opened in

Harry Pollitt's first 4-4-0, class '11' No.694, as built at Gorton in 1895, in Manchester, Sheffield & Lincolnshire Railway livery and with original 3,080 gallon tender, 1895.
(F. Moore/MLS Collection)

694 again, in MS&LR livery, at Louth, c1898.
(MLS Collection)

5697 with Robinson chimney, extended cab roof and 4,000 gallon tender in LNER lined green livery, at Trafford Park, 13 April 1929.
(W. Potter/MLS Collection)

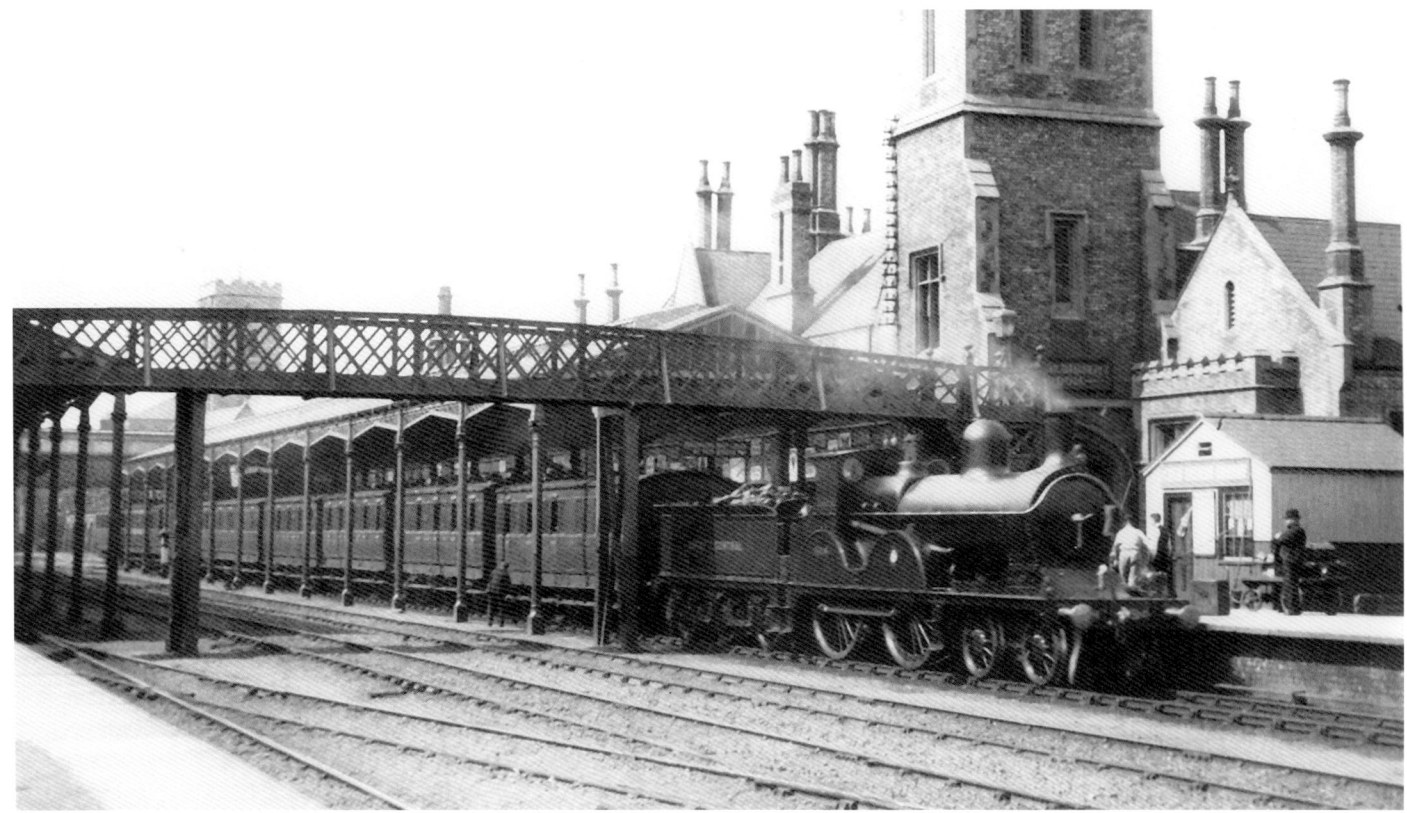

694 at Lincoln with an Eastbound stopping service, 12 August 1899.
(K. Nunn/LCGB/MLS Collection)

699 pilots a C4 Atlantic on a Down express between Leicester and Sheffield, c1912.
(W.A. Brown/MLS Collection)

1899. By 1910, all six were based at Sheffield (Neepsend) working into Lincolnshire (Grimsby) and south as far as Leicester only. They had few opportunities to demonstrate their maximum performance, though there is a record of a run between Grantham and Retford when 698 passed Newark (14¾ miles) in 15 minutes and completed the 33 miles to the Retford stop in 34 minutes 25 seconds with 305 tons, with a top speed of 74mph – timed by the redoubtable Cecil J. Allen.

With the increase of loads during the First World War, members of the class were seen occasionally on expresses as far as Leicester, but usually only as pilots to a GC Atlantic or '11A' 4-4-0. In 1921 the whole class was transferred to Trafford Park for operation over the Cheshire Lines Committee routes to Chester and – less frequently – to Liverpool. As more 'Directors' became available in the immediate pre-Grouping era, D6s took over their CLC work, and they were utilised for local work around Southport, replacing the three Parker D8s and E2 2-4-0s. In 1928, the class was moved back again to Trafford Park to work stopping trains to Wigan, St Helen's, Warrington, Irlam or Chester. This type of work was more suited to tank engines rather than 7ft 4-4-0s and they were withdrawn between 1930 and 1933.

The D6, Henry Pollitt, 1897

The '11A' 4-4-0 class was designed by Harry Pollitt in 1896/7 to provide the main power for the London Extension, due to open in 1899. It was in many respects similar to the class '11' with the following improvements – the provision of piston instead of slide valves and the fitting of larger 4,000 gallon tenders. 268A and 269A were constructed at Gorton in 1897, 270 and 852–861 in 1898 and twenty were built by Beyer Peacock between December 1898 and April 1899 – 862 to 881. Most were superheated between 1912 and 1921, with the remaining five being dealt with by the LNER, the last in 1934. Their dimensions were:

Coupled wheel diameter:	7ft 0in
Bogie wheel diameter:	3ft 6in
Cylinders (2 inside):	18½ x 26in

Stephenson valve gear with 8" piston valves

Boiler pressure:	160lbs psi
Heating surface:	1,101sq ft
Grate area:	19.59sq ft
Axleload:	17 tons
Weight (Engine):	48 tons 11 cwt
(Tender):	43 tons
(Total):	91 tons 11 cwt
Water capacity:	4,000 gallons
Coal capacity:	6 tons
Tractive effort:	14,421lbs

Superheated boilers were fitted to twenty-four engines between 1912 and 1916 and a further four after the war. The LNER renumbered them 5268–5270 and 5852–5881 and fitted the remainder between 1923 and 1929, with just one (5878) not being so equipped until 1934. The superheated boilers were similar to those fitted to the D5s – 170lbs psi, with 20sq ft grate area. The total heating surface was 1,078sq ft for those with 18-element and 1,046sq ft with 15-element superheaters.

The Robinson chimney was replaced by the LNER with

856 with Robinson chimney and original cab as built at Gorton, 1898. (Bob Miller/MLS Collection)

859, with Westinghouse brake pump, extended cab roof and 4,000 gallon tender, at Retford, March 1921. (Real Photos MLS Collection)

'flowerpot' type chimneys when the cracking of the Robinson type became too frequent. Like the D5s, the cab roof was extended from 1912 and Wakefield mechanical lubricators were provided with the superheated boilers. Standard vacuum brakes were fitted in 1902/03, although four – 857, 859, 869 and 876 – had Westinghouse pumps to work air-braked stock coming through from other railways. This equipment was removed in the early 1930s. A few exchanged their 4,000 gallon tenders for the smaller 3,080 gallon ones when they lost their London Extension work to the D9s and 'Directors'.

Their GC lined green livery was replaced initially at the Grouping by LNER lined green, then black with red lining from 1928. The first withdrawals took place in 1930 – 5866 and 5873. Six more went the following year – 5854, 5857, 5858, 5860, 5861 and 5870. There were steady withdrawals from the class through the 1930s until the increased traffic needs of the war called a halt. In 1946, the eight surviving D6s were allocated numbers in the 2100 series though only three, 2101, 2104 and 2106, bore the new numbers. The first withdrawals occurred in 1931 and the last two, 2101 and 2106, in the last month before nationalisation.

D6 5880 in 1928 black livery with red lining, extended cab roof, 'flowerpot' chimney and 4,000 gallon tender, ex-works at Trafford Park, May 1935. (MLS Collection)

5869 at Manchester Central, with Wakefield mechanical lubricator, extended cab, 'flowerpot' chimney and 4,000 gallon tender, March 1939. (W. Potter/MLS Collection)

5871 with 'flowerpot' chimney, extended cab roof and 4,000 gallon tender at Trafford Park, c1945. (MLS Collection)

Operation

The thirty-three locomotives were divided between Gorton and Sheffield during their first few months and worked the Manchester–London expresses which used the trans-Pennine route via Woodford to Sheffield and Grantham before handing over to Great Northern power through to King's Cross. The Great Central's London Extension was opened in March 1899 and many of the D6s were reallocated to Leicester, Woodford and Neasden. The new GC route gained important contracts with the newspaper business and the early morning newspaper trains became known for their speedy schedules and locomotive performance. Initially the 5.15am Marylebone to Manchester, it was retimed to start at 3.45am with new business in October 1899 and retimed again to 2.45am in August 1900. Rumours of fast runs with the Pollitt 7ft 4-4-0s gained publicity with an alleged 208 minute run for the 206 miles, stopping intermediately at Leicester, Nottingham and Sheffield. 268 certainly achieved a time of 220 minutes compared with the best daytime schedule of 245 minutes which itself was not introduced until 1930.

The dominance of the D6s did not last long, however. In 1902, Robinson's '11B' 4-4-0s (LNER D9) began to take over the London work, followed by the Atlantics and, of course, by the first 'Directors' in the final years before the First World War. In the middle of the first decade of the twentieth century, many were allocated to the depots of the former Cheshire Lines Committee (CLC) and by 1909 twenty-nine of the thirty-three

An unidentified Pollitt '11A' 4-4-0 with the royal train at Leicester early in the twentieth century during the reign of Edward VII. (MLS Collection)

were at Trafford Park and Liverpool Brunswick. They not only included the Liverpool–Manchester expresses in their workload but longer distance trains through from Liverpool to Hull. The other four were the Westinghouse braked engines, 857, 859, 869 and 876, which were allocated to Lincoln to work trains such as the Harwich–Liverpool Boat Train which had Great Eastern rolling stock.

During the First World War there was much military work, especially from Liverpool Docks, and D6s were used on troop trains, frequently double-headed. These operated all over the GC system including over the London Extension, but after the war, the D6s were generally confined to the CLC and cross-Pennine routes.

The Lincoln engines in 1921 had moved to Retford (859 and 869 joining 268 already there), the other two to Mexborough. Sixteen were at Trafford Park and eleven at Brunswick with just one at Stockport. The allocation in 1922 changed again. The entire class was split almost equally between Trafford Park and Brunswick. The four Westinghouse fitted engines were reserved for the Harwich Boat trains, but only between Liverpool and Manchester, 857 and 876 at Trafford Park and 859 and 869 at Brunswick.

In 1928, six D6s, 5269, 5865–5867, 5874 and 5875, were transferred to Walton for the Southport services, though shortly afterwards, although remaining at the Southport sub-shed, their parent depot was reinstated as Brunswick. A couple more, 5268 and 5269, were stationed at Stockport to work stopping trains to Liverpool and St Helen's. The Liverpool–Manchester expresses had been accelerated to a 45 minute schedule for the 34 miles, with intermediate stops at Farnworth and Warrington and the D6s had the monopoly of these turns until joined later by the D9s when they in turn were supplanted on the best services by the D10 'Directors'. The *Railway Magazine* of June 1939 printed a number of logs of Manchester–Liverpool expresses timed between the two world wars, and the tables on the next page include a couple of runs with D6s timed by A. Mellor from the Railway Performance Society archive.

881 piloting a Robinson B4 4-6-0, 1101, on what is almost certainly a troop train in the First World War, c1914, somewhere on the London Extension. (Real Photographs/ MLS Collection)

855 on a Cheshire Lines Committee local train for Manchester at Knutsford, c1921. (H. Gordon Tidey/MLS Collection)

		Manchester Central–Liverpool Central					
		5869		5270			
		185 tons		203/215 tons (ex Hull)			
Miles	Location	Times	Speed	Times	Speed		Gradients
0	Manchester Central	00.00		00.00			
1.5	Throstle Nest Junction	03.31		03.28	pws		1/100 F
3.2	Trafford Park	05.29		05.34			1/230 R
6.2	Flixton	08.38	62½	09.23	58½		L
7.5	MP 26½	09.59	53½	10.47	53		1/135 R
9.6	Glazebrook	12.02		12.48	62½		1/135 F
14.2	Padgate Junction	16.28	66	17.06	66		1/240 F
<u>14.7</u>	<u>Warrington</u>	<u>18.43</u>	¾ L	19.05	(18 net)	<u>1 L</u>	
0		00.00	¾ L	00.00		1 L	
2.5	Sankey	04.36	54	04.16	53½		1/180 F, L
4.3	Widnes East Junction	06.43	49½	06.18	50½		1/135 R
<u>6.1</u>	<u>Farnworth</u>	<u>09.08</u>	T	08.25		½ E	1/230 R
0		00.00	T				
2.5	Springfield	04.23	61	11.04	67		1/175 F
5.1	Hunts Cross	07.08	48½	13.27	54		1/185 R

		Manchester Central–Liverpool Central					
		5869			5270		
		185 tons			203/215 tons (ex Hull)		
Miles	Location	Times	Speed		Times	Speed	Gradients
6.8	Garston	08.59			15.09		1/195 F
8.2	Mersey Road	10.16	64½		16.28	70½	1/400 F
9.6	St Michael's	11.40			17.53		
12.2	Liverpool Central	16.12		¼ L	21.45		1 ¼ E

		Liverpool Central–Manchester Central					
		5859			5853		
		196/210 tons			202/215 tons (4pm to Hull)		
Miles	Location	Times	Speed		Times	Speed	Gradients
0	Liverpool Central	00.00	T		00.00	T	
2.6	St Michael's	05.18			05.17		
4	Mersey Road	07.06	52		07.06	51	L
5.4	Garston	08.46			08.50		1/400 R
7.1	Hunts Cross	10.56	46 ½		11.18		1/195 R
					00.00		
9.7	Springfield	13.26	66		04.27	57½	1/185 F
12.2	Farnworth	16.37		1 L	07.33	47½	1/185 R
		00.00		1 L			
1.8	Widnes East Junction	03.40			09.44		1/158 F
5	Sankey Junction	06.50	66		12.48	66/58½	1/158 F, 1/180 R
6.1	Warrington	08.30		½ L	-	via avoiding line	
		00.00		½ L			
1.5	Padgate Junction	03.20			15.00	63½	1/234 F
4.5	Risley Moss	07.03			17.59	56	1/240 R
6.1	Glazebrook	08.49	57½		19.39	62½	L
8.2	Manchester Ship Canal	-	51		-	47	
10.7	Urmiston	13.33	65		24.27	62½	1/135 F, L
12.5	Trafford Park	15.23			26.20		1/180 R
14.2	Throstle Nest E jcn	17.06	sigs 5*		28.04		
15.7	Manchester Central	20.10	(44¾ net)	1¾ L	30.22	(41¾ net)	¼ L

		Liverpool Central–Manchester Central					
		5876			5871		
		222 tons			196/216 tons		
		5.30pm L'pool			7.25pm L'pool		
		6.1.1934			11/4/1936		
Miles	Location	Times	Speed		Times	Speed	Gradients
0	Liverpool Central	00.00		1¾ L	00.00	T	
2.6	St Michael's	04.55			04.50		
4	Mersey Road	06.40	55		6.40	56	L
5.4	Garston	08.30			08.25		1/400 R
7.1	Hunts Cross	10.35	48		10.25	50	1/195 R
8.2	Halewood	12.10	65		12.20	sigs	1/185 F
12.2	Farnworth	16.30		2¾ L	16.35	(16 net)	1 L	1/185 R
		00.00		3 L	00.00		1 L	
	Sankey	07.35	sig stand		05.40		¾ L	
					00.00		1 L	
5	Sankey Junction	11.00			03.00		1/158 F, 1/180 R
6.1	Warrington	13.15	(10¾ net)	4½ L	04.55		1 L	
		00.00		4½ L	00.00		1 L	
1.5	Padgate Junction	03.25	60/50		03.20	63/50	1/234 F, 1/240 R
6.1	Glazebrook	08.20	60		08.20	55	L
9.5	Flixton	12.10	48/58		12.05	50/68	1/135 R, 1/135 F
12.5	Trafford Park	15.40			15.05		1/180 R
14.2	Throstle Nest E jcn	17.30	sigs 5*		17.10		
15.7	Manchester Central	21.30	(19½ net)	7 L	19.00		1 L

As a result of withdrawals of D6s commencing in 1930, former GNR D1s and the later D9s started to augment the remaining D6s at Trafford Park and Brunswick. The increase in holiday traffic during the 1930s saw occasional use of the 7ft 4-4-0s on heavy excursion and summer relief trains to the East Coast resorts. A number of D6s were stored at Gorton during the mid-1930s and in May 1937 5874 and 5880 moved to Immingham to replace the last Parker D7s, followed by 5856 in 1938 and 5865 in 1939. At the start of the Second World War the allocation of the nine remaining D6s was:

Trafford Park:	1
Brunswick:	3
Heaton Mersey:	3
Immingham:	2

Further changes took place during the war years when 5270 and 5855 appeared on the Chester CLC route. They were shedded at Chester and 5859 and 5865 were transferred to Northwich. The last three survivors and the only ones to be renumbered (5855/2101, 5865/2104 and 5874/2106) finished their days at Northwich on stopping trains between Chester and Manchester.

The Great Central 4-4-0s • 55

5871 at Cheadle with an evening excursion to Liverpool and Southport, August 1938.
(W. Potter/MLS Collection)

2101, formerly 5855, leaving Ashley with the 4.31pm Manchester Central–Chester, 3 May 1947.
(J.D. Darby/MLS Collection)

2106 with a Chester bound goods train at Ashley, 3 May 1947. Both 2101 and 2106 were withdrawn at the end of 1947, making the class extinct just before nationalisation.
(J.D. Darby/MLS Collection)

The D7, Thomas Parker, 1887

We go back now to 1887 and the days of the Manchester, Sheffield & Lincoln Railway before its re-emergence as the Great Central Railway and the extension to London. Kitson & Co., locomotive builders, were keen to show an example of their work at the Manchester Exhibition of 1887. The MS&LR had already prepared an E2 2-4-0 for the show, and there was insufficient room for another exhibit. Kitson persuaded the MS&LR management to be represented instead by a new 4-4-0 that Kitson & Co. would build for them, thus killing two birds with one stone. Kitson, in collaboration with the MS&LR's Locomotive Engineer, Thomas Parker, constructed a single-framed 4-4-0 numbered 561 which was delivered to the railway company at the close of the exhibition in November 1887. It was identified as class '2' by the railway company and further examples of the design were not completed until 1890 when Gorton built six, numbered 562–567 and another six, 682–687 in 1891/2.

Parker then placed an order for another twelve with Kitson & Co., which were delivered in 1892 and numbered 700–711. A final batch of six was completed at Gorton in 1894 and classified as '2A' with slight increase in the wheel journals and coiled springs for the driving axles. Their key dimensions were:

Cylinders (2 inside):	18 x 26in
Coupled wheel diameter:	6ft 9in
Bogie wheel diameter:	3ft 6in
Stephenson motion with slide valves	
Boiler pressure:	160lbs psi
Heating surface:	1,278sq ft
Grate area:	18.8sq ft
Axleload:	16 tons
Weight (Engine):	46 tons
(Tender):	37 tons 6 cwt
(Total):	83 tons 6 cwt
Water capacity:	3,080 gallons
Coal capacity:	5 tons
Tractive effort:	14,144lbs

The whole class of thirty-one engines was rebuilt by Robinson with Belpaire boilers between 1909 and 1918. The new boilers had 1,063sq ft of heating surface and a revised grate area of 18.3sq ft. Longer smokeboxes with Robinson chimneys were fitted to 566, 586 and 704 in 1901 and 565, 708, 709 and 711 were similarly equipped extended back supported

The Great Central 4-4-0s • 57

The Kitson built 4-4-0 numbered 561 as built for the 1887 Manchester Exhibition and delivered to the MS&LR in November of that year.
(F. Moore/MLS Collection)

Another view of 561 in traffic, c1888.
(MLS Collection)

685, built in January 1892, seen here with the original round-topped boiler at Grantham shed, 1896.
(Dr.T.F. Budden/MLS Collection)

The 566 as built in 1890 with the original round-topped boiler, but extended roof, at Manchester Central, c1912.
(Bob Miller/MLS Collections)

by extensions to the handrails from 1912. The engines were renumbered 5561–5567, 5682–5693 and 5700–5711 at the Grouping and 'Flowerpot' chimneys replaced the Robinson style from 1924. Two locomotives, 700 in 1902 and 705 in 1903, were equipped with Westinghouse air brakes for working trains of Great Eastern and North Eastern stock. Most were fitted with weather boards to the tender after significant tender-first running was required in their operation in Lincolnshire after the Grouping. Like the Pollitt 4-4-0s, their original GC lined green livery was replaced by LNER lined green after the Grouping and replaced by black with red lining from 1928.

Withdrawal commenced in 1926 with the scrapping of 5564, 5561, the prototype, in 1928 and 5563 in 1929. The remainder were withdrawn in

705, one of the two Westinghouse fitted engines, as newly rebuilt with Belpaire boiler and extended cab roof in February 1911. It has the Robinson chimney, longer smokebox and extended cab roof. (Real Photographs/MLS Collection)

5684 in LNER black livery, with Belpaire boiler, extended cab roof and final form of tender, at New Holland, September 1938. It was withdrawn in June 1939. (MLS Collection)

5704 at New Holland, September 1938
(W. Potter/MLS Collection)

5687 at Gorton Works after withdrawal in August 1935.
(MLS Collection)

the 1930s, the last to go being 5684 and 5704 in 1939.

Operation

Until displaced by the Pollitt Class '11' (LNER D5) in 1895, the Parker 4-4-0s were utilised on the MS&LR's main expresses between Manchester, Sheffield and Grantham, where a GNR locomotive would take over on the King's Cross trains. They would also work the most important trains to Hull. The arrival of the D6s and then the Robinson D9s relegated them to stopping services in the Sheffield, Doncaster, Nottingham and Lincoln areas. After the opening of the London Extension, they could be seen on Sheffield–Leicester stopping services but no further south.

689, built in 1894 at Penistone with a stopping service to Sheffield, c1900. (MLS Collection)

685, built in 1892, at Grantham with a stopping train to Retford, c1900. (Locomotive & General/MLS Collection)

703 with a class '11' (D5) as assistant entering Woodhead Tunnel with a Manchester–Sheffield–London express c1900. (F. Moore/MLS Collection)

708 was sub-shedded at Colwick (GNR) to operate the Nottingham–Sheffield service before the opening of the London Extension, seen here on one of those expresses, c1899. (W.A. Brown/MLS Collection)

707 with a Manchester Central–Liverpool express near Flixton, 1919. (G.M. Shoults/MLS Collection)

The two Westinghouse brake fitted engines were based at Lincoln and Mexborough; at the Grouping, both were at Lincoln. 5700 was withdrawn from there in 1933 and 5705 had its Westinghouse equipment removed in November 1932 and was transferred to New Holland.

In 1921 the allocation was:

Lincoln:	10
Immingham:	10
Retford:	4
Mexborough:	3 (708 & 709 with 710 sub-shedded at Barnsley))
Northwich:	4 (701, 702, 704, 707)

561 and 562 had worked between Southport and Manchester just before the First World War and the 1921 CLC-based engines worked services between Manchester and Chester until around 1930. Various exchanges took place in the 1920s and it was 5565, 5689, 5691 that were withdrawn from Mexborough in 1933 and 5704 transferred to Immingham. By 1933 all the remaining D7s were based in Lincolnshire. Depots in that county that had a D7 allocation at one time or another included Lincoln, Immingham and its sub-shed New Holland, Retford, Louth and Frodingham. Known allocations from the mid-1930s were:

Louth:	5684, 5701, 5703, 5711
Frodingham:	5704, 5708
New Holland:	5684, 5704

(last two years of the class, 1937-1939)

No records exist of their performance in the Railway Performance Society's archives and from the late 1890s they were mainly on stopping services. Little train timing was undertaken or published until the *Railway Magazine* started publication in 1897 and train performance articles began to appear in the early 1900s. It was reported, however, that when working the Manchester–London expresses initially that

5687 leaving Grimsby with a westbound stopping service in the early 1930s. 5687 was withdrawn in 1935. (MLS Collection)

their coal consumption, despite the difficult route, was very economical, as low as 24lb per mile, although admittedly loads were very light.

Preservation

The Great Central Heritage Railway is building a replica of the Thomas Parker/Kitson Class 2, No.567, which was built in 1890, reboilered with a Belpaire boiler in 1918 and became LNER class D7. 5567 was based at Mexborough in January 1930, moved to Immingham in October and was withdrawn in September 1931. The new build 4-4-0 is being constructed at the GCR workshops at Ruddington and by early 2023 the frames had been completed and also on hand were the new cylinder block, a GCR tender and coupling rods. The anticipated cost of the new build locomotive is £500,000. No date has yet been set for its completion.

The D8, Thomas Parker, 1888

Thomas Parker, after the construction of 4-4-0 561 by Kitson for the Manchester Exhibition, designed his own 4-4-0, a development of his own 1887 MS&LR 6D class (LNER E2) 2-4-0. The provision of a four-wheel bogie instead of the single leading axle which was said to be the cause of poor riding. The new July 1888 built 4-4-0 was classified '6DB' and numbered 37. Three months later, Gorton delivered two more, numbered 89 and 400. The Great Central renumbered them 508 (37), 510 (89), and 511 (400). When these numbers were allocated to the D11s in 1919, these three engines were renumbered 508B, 510B and 511B. Parker was clearly impressed by Kitson's 561 and curtailed the construction of any more of his 6DBs and from 1890 built his class '2' and '2As' to the design of the Kitson engine described in the previous paragraphs. Their key dimensions were:

5567 on a local Barnetby–Penistone passenger service at Barnsley Court House in LNER lined green livery, c1925. (MLS Collection)

MS&LR Parker 400 built in October 1888, seen here in the mid-1890s before renumbering 511 by the GCR. (MLS Collection)

Cylinders (2 inside):	18 x 26in
Coupled wheel diameter:	6ft 9½in
Bogie wheel diameter:	3ft 6½in
Stephenson motion with slide valves	
Boiler pressure:	160lbs psi
Heating surface:	1,230sq ft
Grate area:	19sq ft
Axleload:	16 tons 16 cwt
Weight (Engine):	45 tons 19 cwt
(Tender):	31 tons 3 cwt
(Total):	77 tons 2 cwt
Water capacity:	2,600 gallons
Coal capacity:	2¾ tons

They were built with smokebox wingplates which were removed before 1900 and new smokeboxes and Robinson chimneys were fitted in the early 1900s. Their round-topped boilers were replaced by Robinson with Belpaire boilers between 1910 and 1912, as he did the '2' and '2As', and these boilers had 1,063sq ft of heating surface and grate area of 18.3sq ft. At the same time their cab roofs were extended as on the other Pollitt 4-4-0s. The rebuilt engines were 2¼ tons heavier but their maximum axleload had been reduced to 15 tons 18 cwt.

The LNER withdrew 508B and 511B in 1923 before renumbering 510B which became LNER 6415 in December 1924, apparently painted black rather than LNER lined green. 6415 was withdrawn in March 1926.

Parker 6DB 510B (formerly 89), as rebuilt with Belpaire boiler in 1910 and with extended cab roof, seen here shortly after its renumbering in June 1920. (Real photographs/ MLS Collection)

The only D8 to receive LNER livery and number, 6415 (formerly 89, 510B) at Southport, c1925. (MLS Collection)

Operation

Together with the 1887 built Kitson 4-4-0, 561, the three '6DBs' worked the main Manchester–King's Cross trains between Manchester and Grantham until the class '2' 4-4-0s were built in 1890. In 1893 all were shedded at Liverpool Brunswick for the Liverpool Central–Hull services. By 1900, the class '11s' and '11As' were also available in sufficient numbers to remove the 1888 '6DBs' from express work entirely and were transferred to Lincoln and New Holland for local work in Lincolnshire. In 1922, all three were transferred to Southport to operate from there to both Liverpool and Manchester, though 508B and 511B were withdrawn the following year. 510B, now classified as LNER D8 numbered 6415, continued working out of Southport until its demise in March 1926.

The D9, J.G. Robinson, 1901

Robinson's first 4-4-0s appeared from the Sharp Stewart Works just over a year after his appointment as the GCR's Locomotive Engineer in June 1900. Numbered 1013–1017, they were identified as Class '11B' bearing similarities to the Parker Class '11s', reverting to slide instead of piston valves. With traffic booming in the first decade of the twentieth century and the opening of the GCR's London Extension, more express engines were needed and twenty-five had been ordered. The other twenty, 1018–1037, followed in 1902. Their dimensions were:

Cylinders (2 inside):	18½ x 26in
Coupled wheel diameter:	6ft 9in
Bogie wheel diameter:	3ft 6in

Stephenson motion with slide valves

Boiler pressure:	180lbs psi
Heating surface:	1,378sq ft
Grate area:	21sq ft
Axleload:	18½ tons
Weight (Engine):	53½ tons
(Tender):	48 tons 6 cwt
(Total):	101 tons 16 cwt
Water capacity:	4,000 gallons
Coal capacity:	6 tons
Tractive effort:	16,850lbs

The reason for the reversion to slide valves was that the '11As' piston

510B just before renumbering 6415 entering Allerton Sidings with the stock of a Southport–Hunts Cross train, 5 August 1924. (H.A. White/MLS Collection)

valves gave trouble and several different types were tried before satisfaction was reached. By the time of the design work for the '11B', solutions had not yet been found. Five more '11Bs' were built by Sharp Stewart in 1903 and were numbered 1038–1042 and a final ten were constructed by Vulcan Foundry and received the numbers 104–113. The Sharp Stewart built locomotives were originally coupled with 3,250 gallon tenders but the Vulcan engines had 4,000 gallon tenders and water pick-up scoops as the company were laying water troughs and all the Sharp Stewart engines had received 4,000 gallon tenders off 0-6-0 and 0-8-0 goods engines by 1906.

In 1907, Robinson fitted a larger diameter boiler to 104 and 110, increasing the grate area to 26sq ft – identical to the grate areas of the successful Atlantics being built at Gorton at the same time. The cylinder diameter was increased to 19in. The overall heating surface was raised to 1,626sq ft and weight of the engine was over a ton heavier at 54 tons 16 cwt. The tractive effort was raised to 17,729lbs. These two rebuilds were classified as ''11C'. 104 was named *Queen Alexandra* in 1907 and 110 *King George V* in 1911 on his accession to the throne. 104 retained these design features until 1923 when it was rebuilt as a superheated '11D'. 110 exchanged its boiler for a saturated '11B' boiler in 1918, but the large boiler was repaired and fitted to 113 which it retained until 1923 when it received a superheated boiler. '11B' 1021 was named *Queen Mary* in 1913. 1014 had been named *Sir Alexander* when built in 1902, after the company's Chairman, but lost it in 1913 when the first of the D10 'Directors' was named *Sir Alexander Henderson*.

'11B' 1026 was fitted with the larger boiler like 104 and 110 but retained the original heating surface and grate area. It was given later type piston valves and was classified as '11D'. Similar superheated boilers (some with 18 elements and some with 22 elements) were constructed to rebuild the '11Bs' from 1913, 1021 being the first to receive the 5ft diameter superheated boiler with 24 elements and a new cylinder block with 20in diameter by 26in stroke and 10in piston valves. However, Gorton later standardised the 22-element superheater, though some engines exchanged their boilers with 18-element superheaters as fitted in 1911 to the A5 4-6-2 tank engines. They were fitted with mechanical lubricators at the same time. By the time of the Grouping, most of the class had reverted to 19in diameter cylinders and 21sq ft grate area. It is thought

An official photograph of the last '11B' to be completed, 113, in June 1904. (MLS Collection)

The Great Central 4-4-0s • 69

Railways received the mixed traffic livery, black with LNWR style lining – 62313, 62317 and 62332.

Withdrawal of three was planned in 1939 but was halted as a result of the need for extra locomotives during the 1939–45 hostilities. A limited programme of withdrawals restarted in 1942 and six D9s had been withdrawn by the end of the war (5110, 5113, 6020, 6022, 6028, 6042). Thirty-four remained and in 1946 were allocated the numbers 2300–2333, though six (2310, 2316, 2320, 2323, 2326 and 2327) did not survive long enough to receive

Vulcan built 104 *Queen Alexandra* as rebuilt with a larger boiler as Class '11C' in 1907. (Real Photographs/MLS Collection)

that the larger grate area cramped the design of the ashpan. The rebuilding was almost completed by the Great Central, just seven left for the LNER to reboiler, the last, 6042 (ex-1042) in 1927. The revised dimensions of the superheated '11Ds' were:

Cylinders (2 inside)   19 x 26in
Heating surface: 1,524sq ft
(with 18 element superheater of 145sq ft)
   1,458sq ft
(with 22 element superheater of 178sq ft)
Grate area: 21.17sq ft
Weight (Engine):   55 tons 14 cwt
Tractive effort: 17,729lbs

The engines were reclassified as Class D9 and renumbered after the Grouping, adding 5000 to their GC numbers becoming 5104–5113 and 6013–6042. The locomotives exchanged their GC lined green for LNER lined green in 1924 and black with red lining in 1928. Three of the engines that passed to British

Left: **1021 was** named *Queen Mary* in 1913. It is seen here at Manchester London Road shortly afterwards. (P.F. Cook/MLS Collection)

Below: **113 received** the larger boiler off 110 in 1918 and is seen here shortly after that exchange. The longer firebox is very evident in this photograph. (Bob Miller/MLS Collections)

70 • LONDON & NORTH EASTERN RAILWAY 4-4-0 TENDER LOCOMOTIVES

6034 in the LNER lined green livery which this engine bore between 1924 and 1928. Note the GC style oval new numberplate on the cabside. (Bob Miller/MLS Collection)

5110, *King George V,* the D9 whose larger boiler was transferred to 113 in 1918, seen here in LNER lined green livery, c1926. (W. Leslie Good/MLS Collection)

The Great Central 4-4-0s • 71

5104, former GC '11C' 104, rebuilt with superheated boiler as Class '11D' in 1923, seen here in post 1928 LNER black livery as in standard D9 superheated form, 5104 *Queen Alexandra,* at Liverpool Brunswick CLC shed, c1937. The number has been painted low on the cabside to avoid removing the splasher brass beading that covers the area where the number should have been painted. Note the smaller chimney and reduced height dome. (Bob Miller/MLS Collections)

2318, formerly GC 1034, LNER 6034, rebuilt as Class '11D' in 1914, at Manchester Central after renumbering in February 1949. It was withdrawn in November of that year. (P. Ward/MLS Collection)

62317 (built as 1033 in 1902), one of the D9s painted in BR mixed traffic livery in September 1948. It was withdrawn in July 1949. (Robert Fysh/MLS Collections)

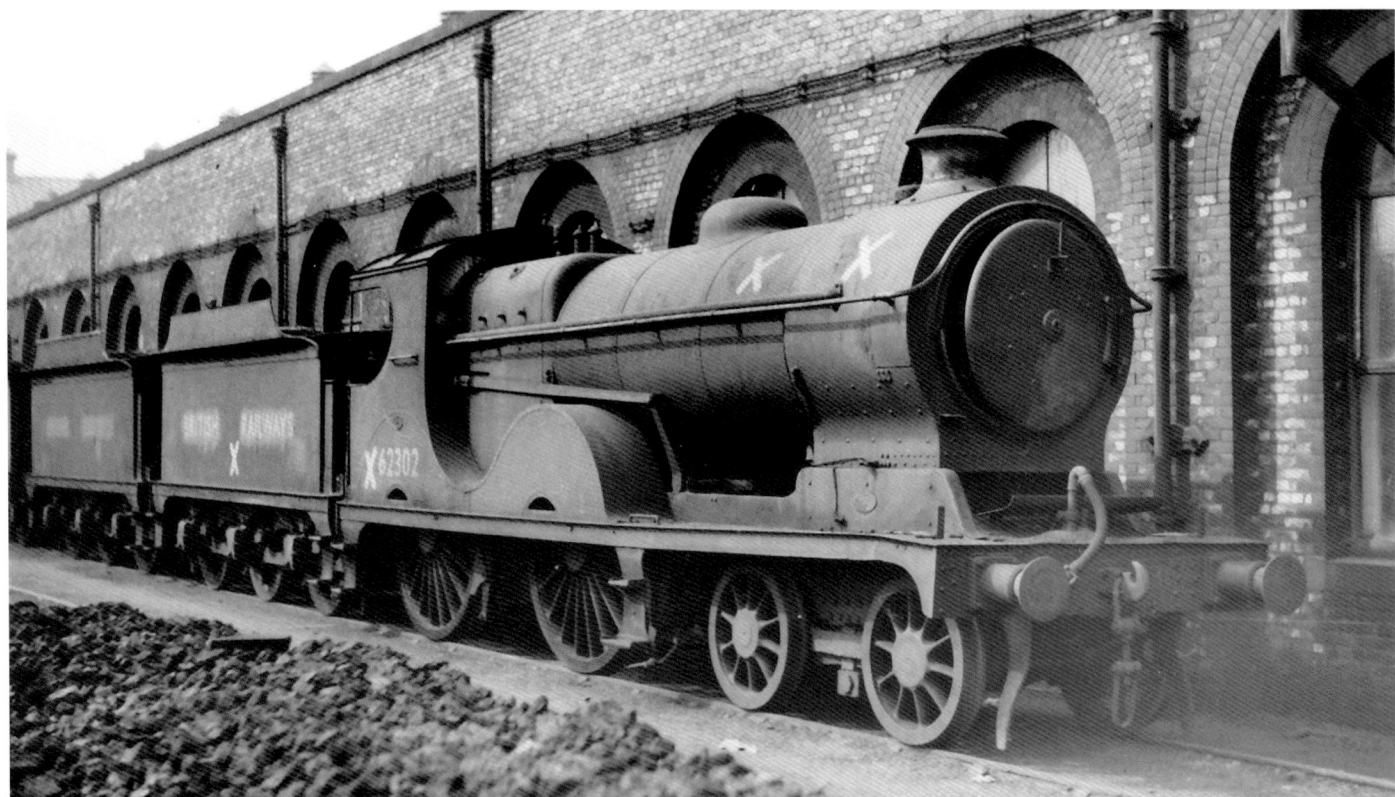

62302, built as 1015 in 1901, seen at Gorton after its withdrawal in February, 22 April 1950. (MLS Collection)

their new 60,000 numbers. Twenty-six were transferred to British Railways in January 1948 and the last one, 62305 (ex-1018/6018) was withdrawn in July 1950 from Trafford Park shed.

Operation

The '11Bs' were initially allocated to Gorton, Leicester and Neasden, replacing the Pollitt 4-4-0s (the '11As') on the most important expresses on the London Extension until they themselves were displaced by the Robinson Atlantics and the 1913 'Directors'. In the early days, loads were light – no more than three or four coaches, weighing only 100-120 tons – but the schedules were fast, especially after 1905. I cannot find any logs from the very early days, though

Vulcan built 111 at Leicester on the 8.45am Marylebone–Manchester, April 1911. (G.M. Shoults/MLS Collection)

111 again on the 5pm Leicester–Marylebone stopping train leaving Leicester, 4 June 1911. (G.M. Shoults/MLS Collection)

Large boilered 104 *Queen Alexandra* after its transfer to Woodford, with a horsebox special for the Great Western (possibly Newbury), c1912.
(W.A. Brown/MLS Collection)

112 passing Neasden with the 4pm express to Leicester, c 1910.
(MLS Collection)

1024 with a London–Manchester express leaving Nottingham, c1910. (F. Moore/MLS Collection)

1029 on the 2.15pm Manchester–Marylebone express near Whetstone, c1910 (G.M. Shoults/MLS Collection)

1014 with the 9.25am from Manchester entering Hull Paragon station, 22 June 1920. The rolling stock seems below par for an express as late as that.
(K. Nunn/LCGB/MLS Collections)

1040 with a CLC express near Hunts Cross East Junction, c1920.
(H. Gordon Tidey/MLS Collection)

it is reported that 1040 covered the 126½ miles to Nottingham in 123 minutes and worked through to Manchester, 206 miles, in 219 minutes, compared with the best 1930s schedule of 245 minutes, admittedly with heavier loads and more stops.

After their London express work petered out before the First World War, most were based in the Sheffield area, their work being on the Manchester–Hull axis and also to Woodford and Banbury with trains to Great Western destinations. The two large boilered '11Cs' after regular work between Manchester and Hull, were transferred to Woodford for the latter type of duty.

At the Grouping their allocation was:

Sheffield:	18
Annesley:	9
Immingham }	
Lincoln }:	11
Retford }	
Brunswick:	1
Mexborough:	1

The Lincoln engines worked the Harwich–Liverpool boat train between Lincoln and Sheffield until 1927, when former GE B12 4-6-0s took over right through from Ipswich to Manchester, but they still worked the York portion. They worked local trains to Retford, Sheffield, Doncaster and New Holland. The Sheffield engines worked to York, Hull, Cleethorpes, Lincoln, Leicester, Liverpool and Manchester. However, a number were replaced by ex-GNR Ivatt Atlantics which were allocated

Sheffield Victoria–Leicester Central
5105, 9 chs
1pm Sheffield–Marylebone
27.4.1928

Miles	Location	Times	Speed	Gradient
0	Sheffield Victoria	00.00		
2	Darnall	05.40	pws	1/144 R
4.9	Woodhouse	10.15		1/137 F
7.4	Killamarsh	14.15	pws	1/176 R
10	Eckington	18.25	pws	
12	Staveley	21.45		1/176 R
17.7	Heath	33.05	30.3 (avg)	1/100 R
20.2	Pilsley	36.55	39.5 (avg)	1/300 R
21.6	Tibshelf	38.30	60	1/132 F
26.4	Kirkby Bentinck	43.35	57	1/132 F, L
32.3	Hucknall	49.35	59 easy	1/132 F
36.5	New Basford	53.30	60	1/130 F, 1/330 R, 1/130 F
38.1	Nottingham Victoria	56.25		
0		00.00		
1	Arkwright Street	02.30		
4.4	Ruddington	06.50		L
9	East Leake	12.00	53 (avg)	1/176 R
13.6	Loughborough	16.50	57.5 (avg)	1/176 F
15.6	Quorn	18.55		1/176 R
18.4	Rothley	22.05	53 (avg)	1/264 R
21.1	Belgrave	25.10	52.2 (avg)	1/176 R
23.5	Leicester	27.55	(changed engines to 4-4-2 6085)	

	Banbury–Nottingham Victoria 5105, 4 chs 10.50am Oxford–Sheffield 16.6.1931			
Miles	Location	Times	Speed	Gradients
0	Banbury	00.00		
1.2	Banbury Junction	03.05		
3.4	Charwelton Road	06.30	39	
5.4	MP 4	09.55	33	Summit
7.2	Eydon Road	12.15	53	
9.4	Culworth Junction	14.50	45	1/176 R
11.2	Woodford	17.00	47	
13.6	Charwelton	20.00	50	1/176 R
17.3	Staverton Road	23.35	72	1/176 F
20.6	Braunston	26.20	78	1/176 F
25.4	Rugby	31.00		
0		00.00		
3.4	Shawell	0520	50/47	1/176 R
6.8	Lutterworth	09.10	56/50	1/176 F, 1/176 R
10.7	Ashby	13.35	56	1/176 F
15.2	Whetstone	17.40	69/72	1/176 F
19.8	Leicester	22.25		
0		00.00		
2.4	Belgrave	04.30		1/176 R
5	Rothley	07.55		1/176 F
7.8	Quorn	10.40	64	1/264 F
10	Loughborough	13.00		
0		00.00		
3	Barnston	05.25	40	1/176 R
4.5	East Leake	07.15	56	1/330 F
5.5	Rushcliffe	08.10	64	1/176 F
7	Gotham	09.30	75	1/176 F
9.2	Ruddington	11.20	72	L
13.6	Nottingham Victoria	16.20		

primarily the turns to Manchester and Leicester. Then, in the late 1920s, Robinson 4-6-0s took over the Manchester–Sheffield–Hull/ Cleethorpes trains, though the D9s retained their York–Sheffield– Woodford turns including the overnight Penzance–Aberdeen through coaches between Sheffield and York. Gerald Aston timed a couple of typical services of this period with 5105, a Sheffield engine.

Between the war years, the Immingham engines monopolised the New Holland/Cleethorpes–Doncaster trains and later in the thirties a Grimsby–Peterborough express connected there for London. It did a fill-in trip up the East Coast line with a stopper to Hitchin and back before returning to Grimsby in the evening, although C4 Atlantics took over this diagram a year or so later. 6021 and 6034 went to Leicester in 1933 to act as pilot engines to the Atlantics on the heaviest trains until the B17 'Footballs' made their appearance in 1937. The former GNR shed at Colwick received 6022, 6025, 6035, 6037–6039 from Annesley in 1928 for services starting from Nottingham. Annesley retained three for Mansfield trains and station pilot duty (which 6016 covered for a number of years). As the B17s displaced the D10 and D11 'Directors' on the main line to London caused a number of D9s to be transferred to the CLC shed at Brunswick and Trafford Park, displacing in turn the Pollitt D6s. By 1939 there were six D9s at Brunswick and five at Trafford Park.

1037 at Nottingham Victoria with a train of Great Western stock for Banbury and Swindon. Note that 1037 has received the LNER lined green livery but has not yet been renumbered, c1923. (MLS Collection)

6016 leaving Sheffield Victoria with the 3.35pm slow train to Leicester, 3 September 1932. (MLS Collection)

A number of logs exist from pre-war *Railway Magazines* and the Railway Performance Society's archives of their running in the 1930s on the Manchester–Liverpool expresses. A selection in both directions is given below.

The 74mph for the run with 5110 was exceptional and the other two runs are more typical of the performance on this route. The December run with 5104 was run in appalling weather conditions, sleet from Manchester to Farnworth and driving hail afterwards.

		Manchester Central–Liverpool Central						
		5110 *King George V* 175/185 tons		**5104** *Queen Alexandra* 7 chs, 215 tons 2.45pm M'chester 5.12.1936		**6020** 8 chs, 192/210 tons 2.45pm M'chester 31.7.1937		
Miles	Location	Times	Speed	Times	Speed	Times	Speed	Gradients
0	Manchester Central	00.00	T	00.00	1 L	00.00	5 ½ L	
1.5	Throstle Nest East Jn	03.32		03.35		03.21		1/132 F
3.2	Trafford Park	05.30		05.16	54	05.21	53/50	
6.2	Flixton	08.37	65	08.26	64	08.29	62	1/180F, L
7.5	MP 26½ (Ship Canal)	09.51	58½	-	54	-	53	1/135 R
9.6	Glazebrook	11.45	68½	11.31	63	11.37	63/67	1/135 F, L
14.2	Padgate Junction	15.46	74	16.30	56	16.13	63	1/240 F, L
14.7	Warrington	17.55	T	18.28	1½ L	18.06	5¾ L	
0		00.00		00.00	¾ L	00.00	8½ L	
2.5	Sankey	04.34	53½	04.28	53	04.33	51	1/180 F, L
4.3	Widnes East Junction	06.35	50	06.31	48	06.42	45	1/158 R
6.1	Farnworth	09.07	1 E	09.03	T	09.17	8 L	
0		00.00		00.00	2 L	00.00	8 ¾ L	
2.5	Springfield	03.56	61½	-	50½	-	55	1/185 F
5.1	Hunts Cross	06.48	52	07.33	47	07.37	45	1/185 R
6.8	Garston	08.41		09.22	58	09.24	62	1/195 F
8.2	Mersey Road	10.05	57 easy	10.57	62	10.54	64	
9.6	St Michael's	11.44		12.23	53	12.12	61	
12.2	Liverpool Central	16.09	T	16.38	2¾ L	16.18	9 L	

		Liverpool Central to Manchester Central								
		6033			6019			6020		
		259/280 tons			7 chs, 215 tons			8 chs, 280 tons		
					4pm Liverpool–Hull			8.30am Liverpool – Manchester		
					30.10.1935			17.2.1936		
Miles	Location	Times	Speed		Times	Speed		Times	Speed	Gradients
0	Liverpool Central	00.00		T	00.00		¾ L	00.00		¾ E	1/101 F
2.6	St Michael's	04.50			06.01	43		05.46		1/195 R
4	Mersey Road	06.30	54		07.46	51½		07.32	49	
5.4	Garston	08.03	51½		09.34	48		09.19	47	1/185 R
7.1	Hunts Cross	09.57			11.57		½ L	11.28	41 ½	
					00.00		½ L		sigs	
9.7	Springfield	12.34	67		-	44		-	pws	1/185 F
12.2	Farnworth	15.38		T	07.16	58/51		20.09	3½ L	
0		00.00		T				00.00	3½ L	
1.8	Widnes East Jn	03.05			09.17	55		03.39		
5	Sankey	-	61		10.58	60		05.35	59	1/185 F
6.1	Warrington	08.52		¼ E	-	(avoiding line)		08.51	3¼ L	
0			00.00		T			00.00		4¾ L
1.5	Padgate Junction	03.05			14.29	50		04.10		1/240 F
4.5	Risley Moss	06.51	55		-	57		-	54	1/240 R
6.1	Glazebrook	08.42			19.40	pws		08.53	54	8.2
	MP 26½ (Ship Canal)	-	46		-			-	48	1/135 R
10.7	Urmston	13.43	63½		27.33	56		14.04	54	L
12.5	Trafford Park	15.42			29.28	55/60		18.29	pws	
14.2	Throstle Nest E Jn	17.35	sigs		-			-		
15.7	Manchester Central	21.09		2 L	33.50		4½ L	21.38	6½ L	

Other runs documented are similar to the majority of those above – the D9s struggled to keep time if there were any p-way slacks or signal checks. In March 1936, 6019 with eight coaches just about managed to keep the schedule of the 4.30pm Manchester including a slight p-way slowing at Trafford Park, maximum speeds of 67 at Glazebrook and 65 at Garston. In May 1939, 5111 with six coaches, 144 tons tare, ran from Manchester to Warrington in 18 minutes 15 seconds, maximum 63½mph, Warrington to Farnworth in 9 mins 22 seconds and Farnworth–Liverpool in 17 min 20 secs, top speed 60mph, unable to recover the 3½ minute late start. 6037 with a heavier load of 9 coaches (270 tons)

5106 at Retford with a Manchester–Cleethorpes train, 25 July 1936. (MLS Collection)

on the 4.30pm Manchester–Liverpool also in the some month, struggled losing a minute to Warrington, max speed 60mph, lost three minutes between Warrington and Farnworth because of a p-way slack and lost another minute between Farnworth and Liverpool, arriving 7 minutes late.

Other runs logged in the 1930s included a couple of Sheffield–York runs. John Wrottesley travelled on the 7.40am from Swansea to York in August 1933, with Churchward mogul 6337 all the way from Swansea to Banbury, GC Atlantic 5262 to Leicester, D10 5509 to Sheffield and D9 6040 with six coaches from Sheffield to York. In January 1936, 5111 had five coaches (133 tons) on a York–Sheffield run, leaving York 2½ minutes late, with a maximum speed of 68mph between Copmanthorpe and Church Fenton and 67 at Moorthorpe, but signal checks meant a 7 minute late arrival in Sheffield, the 46.8 miles having taken 64¼ minutes, much through coal mining subsidence areas. Albert Mellor logged Sheffield's 6017 on the 5.03pm Worksop–Sheffield in 1934, but the last few miles from Woodhouse into Sheffield were plagued by signal checks and 7 minutes were lost in half hour run.

In 1934 one or two began to be sent to East Anglia. 6022 to Ipswich, 6031 to Norwich, 6024 to March then Peterborough East, 6018 to King's Lynn. By 1935 the following locomotives were allocated to sheds in the former Great Eastern territory:

King's Lynn: 5109, 6015, 6018, 6035
Peterborough East: 5113, 6024, 6025, 6027, 6031, 6040-6042

They worked to Cromer and Sheringham, Hunstanton, Cambridge and Peterborough, often turn and turn about with ex-GE D15 'Claud Hamiltons'. One of the Peterborough engines was diagrammed to the night mail train to Liverpool Street via Ely and Ipswich, returning the following night direct from Liverpool Street to Cambridge, Ely and Peterborough. The Railway Performance Society's archives have three logs of the return mail train, the 10.10pm from Liverpool Street but for some reason the recorder, Peter Proud, only recorded the times as far as Broxbourne where he lived – although it shows Broxbourne as a passing time. 6016 with twelve coaches, 325 tons, on 19 November 1938 passed Broxbourne in 26 minutes 7 seconds for the 17.1 miles, maximum speed over the gently rising gradients of 50mph. 5113 on 8 December 1938 had 345 tons, but took 30 minutes 37 seconds, with 51mph maximum before a p-way slack at Ponders End. 6019 on 8 April 1939 had a load of 420 tons and struggled with nothing over 46mph, taking 28¾ minutes to pass Broxbourne.

The D9s gained more work in the area when the LNER took over the Midland & Great Northern Railway in 1936. The allocation to Peterborough East had risen to eleven by 1937 and further D9s went to South Lynn. The most prestigious D9 M&GN turns were for South Lynn engines on expresses from the Norfolk Coast to Peterborough and Leicester. 6038 and 6042 moved on to Yarmouth Beach.

During the war years there were further reallocations. In January 1943, the distribution was:

Brunswick Liverpool: 9
Trafford Park: 4
Stockport: 2
Mexborough: 3
Barnsley: 1
Staveley: 1
Immingham: 1
March: 6
New England: 4
South Lynn: 2
Yarmouth Beach: 2

More movement took place during the last years of the war, with the D9s being concentrated on the former CLC lines and East Anglia. The schedules were decelerated from 45 to 55 minutes for the Manchester–Liverpool run and even that could provide problems with increased loads and deteriorating condition of the engines through lack of maintenance. For instance, 5104 in February 1943 had ten coaches, 312 tons on the 4.15pm Liverpool–Hull but took 58 minutes with a maximum speed of 54mph, most of the journey being completed in the 40s. By 1946 all the remaining D9s had returned to the CLC lines apart from 6018 and 6040, which spent a couple of months at Stratford at the end of 1945. On 12 January 1946, the Rev. R.S. Haines timed 6018 at the head of a Liverpool Street–Cambridge train that left Bishop's Stortford theoretically at 7.43pm although it was already very late. It was loaded to ten vehicles and managed to drop a further 10 minutes (net) on the 35 minute schedule for the stopping train's 15.7 miles. After its arrival at Whittlesea where the Rev. Haines left the train, some 49 minutes late, he commented, 'Neither a D9 or K2 should be put on any but the lightest or easiest timed passenger trains.' Withdrawals had reduced the class to twenty-seven members by January 1947, the allocation being:

Brunswick Liverpool: 16
Trafford Park: 7
Stockport: 1
Walton: 1
Northwich: 2

Their maintenance had been neglected throughout the war and their post-war condition was such that they continued to struggle with the hourly Manchester expresses even though the wartime schedule of 55 minutes still applied. 2307 (ex-6021) on the 7.30pm Manchester–Liverpool relief train in August 1946 just kept time with a maximum speed of 59mph before the climb to Hunts Cross surmounted at 43mph. 2308 (ex-6023) in March 1947 on the 2.30pm Liverpool managed to keep time as far as Warrington, max speed 59mph, but dropped over 3 minutes between Warrington and Manchester because of a 30mph p-way slack, max speed on this stretch being only 48mph. 62332 (ex-5111), which was – at least externally – in better condition than most of the rest of the class, hauled the 7.10am Liverpool Lime Street–Manchester in January 1949, dropped 2 minutes to Warrington, max speed 52mph, had a p-way slack at Sankey and arrived in Manchester still 2 minutes down with just 51½mph at Urmston. The Cheshire Lines became the

responsibility of the London Midland Region in 1948 and they were soon replaced by former LMS power, the last survivor being 62305 (former 6018) which was withdrawn from Trafford Park in July 1950.

The D9s were a significant improvement on the earlier Great Central 4-4-0s and enjoyed their heyday on the London Extension in the first decade of the twentieth century until displaced first by the Atlantics and then by the 'Directors'. After that they undertook secondary duties, unassumingly, with their association greatest with the former CLC routes, especially the Manchester Central–Liverpool Central expresses. In the war years and the immediate post-war period, they were neglected and struggled to carry out their allotted tasks, 'wheezing out' their remaining days on stopping trains in Cheshire.

6013 and 0-6-0 4376 at Manchester Central, 1 October 1946. 6013 is at the head of an express for Liverpool.
(H.C. Casserley/MLS Collection)

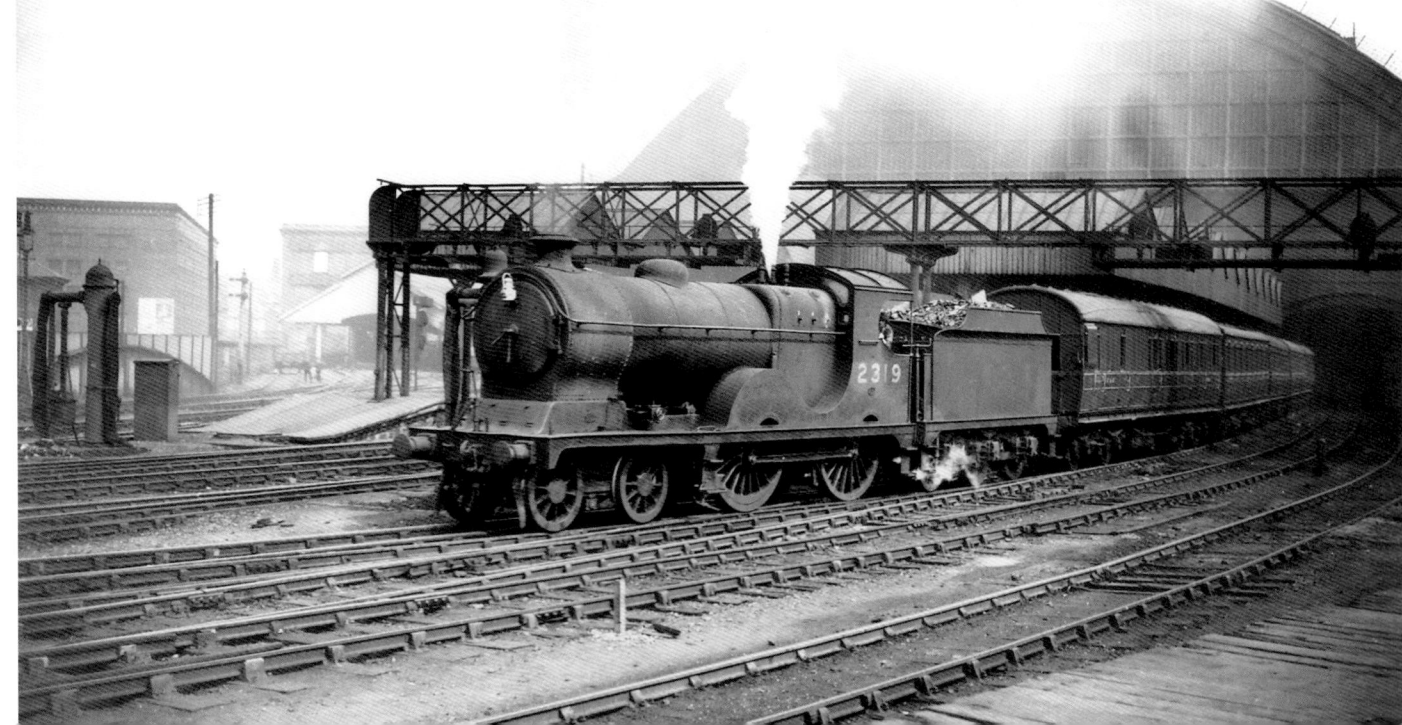

2319 leaving Manchester Central with a slow train for Liverpool, 30 August 1947.
(J.D. Darby/MLS Collection)

2322 on the 6.44pm Chester–Manchester Central leaving Ashby, 11 June 1947.
(J.D. Darby/MLS Collection)

Below left: **2305 double-heads** a tender-first 0-6-0 on a goods train between Hale and Ashley, 9 August 1947.
(J.D. Darby/MLS Collection)

Below right: **2302 at Cheadle** on the 3.23pm Stockport–Liverpool Central stopping train, 10 October 1948.
(MLS Collection)

62332 and 0-6-0 65179 at Skelton Junction, 3 February 1949.
(N.H. Spilsbury/MLS Collection)

The D9 that was kept in reasonable external condition in BR mixed traffic livery, 62332 again, Leaving Sefton with a Southport train, 11 May 1949.
(MLS Collection)

62312 at Manchester Central with an express headcode excursion, 17 April 1949. (MLS Collection)

A run down and filthy 62313 passing Wavertree Junction with a Manchester–Liverpool Lime Street stopping train, 15 April 1949. (MLS Collection)

62305 and an LMS Stanier 3MT 2-6-2T at Manchester Central, 13 June 1949.
(H.C. Casserley/MLS Collection)

62300 enters Manchester Central with the 10.30am express from Liverpool Central, 23 April 1949.
(J.D. Darby/MLS Collection)

Tender-first 62333 draws into Cheadle with a stopping train to Manchester Central, 1949. (MLS Collection)

The D10, J.G. Robinson, 1913

Passenger traffic built up on the main Great Central routes in the first decade of the twentieth century and Robinson's '11Bs', though competent for the initial light loads on the London Extension, were soon considered too small. The Atlantics, later LNER class C4, and the four C5 compounds took over very successfully and by the end of 1906 there were thirty-one of them which continued on main line work until the late 1930s when the B17 'Footballers' appeared on the GC main lines. Other railways, notably the Great Western, were turning out extremely impressive 4-6-0s and Robinson designed the GCR class 1 (LNER B2) 4-6-0s known as the 'Sir Sam Fays'. They were disappointing, their performance not meeting expectations, with poor steaming and high coal consumption, indicating in particular problems in the firebox/ashpan design constrained somewhat by the 4-6-0 layout. If Robinson were to acknowledge this failure – which he did – and needed more front line passenger engines, he could have constructed more Atlantics. What he did was at first appearance a backward step. He designed another 4-4-0 but with a real balance between boiler and cylinder size and improvements in the ashpan design made possible by the room, 10ft between the coupled axles instead of 8ft 3in between the two rear coupled axles for the 'Sir Sam Fays' provided by the 4-4-0 wheel arrangement, his masterpiece, the class 11E, or as they became known, the 'Directors', later LNER class D10.

Ten locomotives were constructed at Gorton Works in 1913 and were numbered 429–438. They were turned out in the full Great Central passenger livery of lined green and named after the company's directors at the time:

429 *Sir Alexander Henderson*
430 *Purdon Viccars*
431 *Edwin A. Beazley*
432 *Sir Edward Fraser*
433 *Walter Burgh Gair*
434 *The Earl of Kerry*
435 *Sir Clement Royds*
436 *Sir Berkeley Sheffield*
437 *Charles Stuart Wortley*
438 *Worsley Taylor*

Sir Alexander Henderson was the company's Chairman and '11B' 1014 was already named *Sir Alexander* but gave it up to the prototype of the new class, with his surname added. He and most of the others remained directors until the Grouping and four of them became directors of the LNER. The dimensions of the new 4-4-0s were:

Cylinders (2 inside):	20 x 26in
Coupled wheel diameter:	6ft 9in
Bogie wheel diameter:	3ft 6in
Stephenson motion with 10in piston valves	
Boiler pressure:	180lbs psi
Heating surface:	1,963sq ft (of which superheater 304sq ft)
Grate area:	26.5sq ft
Axleload:	19 tons 16 cwt
Weight (Engine):	61 tons
(Tender):	48 tons 6 cwt
(Total):	109 tons 6 cwt
Water capacity:	4,000 gallons
Coal capacity:	6 tons
Tractive effort:	19,644lbs

As indicated earlier, part of the success of the '11Es' was the fitting of the 8ft 6in length firebox with a deeply sloping grate enabling improved ashpan and draughting design. The boiler tubes were shorter in the 12ft 3in boiler barrel, 5ft less than the 4-6-0s, and were coupled with 20in diameter cylinders which balanced the boiler steam raising capacity rather than the larger 21½in of the larger engines.

Robinson designed a new boiler in 1914 for his 2-6-4 tank engines (the LNER L3 class) and these were used later for the further six 'Directors' of class '11F' (the D11s described in the next section of this book). This type of boiler was later fitted to the '11Es' and by the Grouping all were so equipped. The number of small tubes was reduced from 175 to 157 and superheaters with short loop elements reduced the overall heating surface to 1,752sq ft of which the superheater surface was only 209 instead of 304sq ft. The engines were provided with the now standard Wakefield mechanical lubricator and Ramsbottom pattern safety valves.

The 'Directors' were allocated the numbers 5429–5438 at the Grouping and given the LNER lined green livery although they too suffered the change to black with red lining that the company adopted for many of its secondary passenger engines as an economy measure in 1928. Some name changes had taken place in the latter years of the Great Central. The introduction of the GC 1A 4-6-0s in 1917 as the new company prestige engines caused the name of the Chairman (who became Lord Faringdon in 1916) to be applied to the first of

429 *Sir Alexander Henderson* as built in August 1913.
(F. Moore/MLS Collection)

The Great Central 4-4-0s • 91

430 *Purdon Viccars* at Gorton shortly after construction, 1913.
(MLS Collection)

436 *Sir Berkeley Sheffield* seen running in after its construction at Gorton in November 1913.
(Bob Miller/MLS Collections)

the new class and 429 was then renamed *Sir Douglas Haig*. When he received an Earldom another of the '1A' engines, 1166, received that name, 429 was renamed again as *Prince Henry* who became known later by the public as the Duke of Gloucester. One of the other directors, Charles Stuart-Wortley, made a Baronet in 1916, gave his name to the '1A' 1168, thus releasing 437 to become *Prince George* (the future Duke of Kent).

The height above rail level with the Robinson chimney was over 13ft 1in which did not, like many other GC engines, conform to the LNER Composite Loading Gauge. Some reduction in the height of boiler mountings took place later as the LNER traffic management required to use the engines in other Motive Power Divisions. A number of LNER locomotives in the 1930s were fitted with Trofinoff Automatic By-pass piston valves. The purpose was to avoid compression of air in the cylinders when coasting. There were long stretches of 1 in 176 gradients on the GC main line and steep winding gradients on the Manchester–Sheffield route where coasting was common and 5431 was chosen as one of the experimental engines in January 1935. Whereas drivers normally kept the regular slightly open when coasting, instructions were issued for 5431 that the regulator should be fully closed when coasting was appropriate. The experiment was successful and 5429, 5430 and 5435 were similarly equipped in 1937 and 5432 in 1938. It is probable that the rest of the class would have been so fitted but the outbreak of the war stymied further developments.

All members of the class survived the war and were renumbered under the 1946 scheme 2650–2659. The red lining had been discontinued during the war and until nationalisation the locomotives were plain black. In 1948, as part of British Railways stock, they were again renumbered as 62650–62659 and they received the BR mixed traffic livery of black with LNWR style lining. Some then lapsed back to plain black although 62652–62654, 62656 and 62658 still wore the mixed traffic livery when they were withdrawn. 62651 and 62657 were the first to be withdrawn in March 1953 with a couple more in August. The last three survivors were 62656 withdrawn in January 1955, 62658 in August and 62653 in October.

Three photos showing the livery and naming change for 437, seen here as built in 1913 as *Charles Stuart-Wortley*. (MLS Collection)

437 renamed *Prince George*' a year later in LNER lined green livery and identified as 437c (identifying a loco of the former Great Central Railway) before being allocated its new number of 5437 later in November, taken at the same location and by the same photographer as the previous photograph, 18 June 1924. (W. Potter/MLS Collection)

437 renumbered 5437 in November 1924 at Gorton early the following year. (Real Photographs/MLS Collection)

5435 *Sir Clement Royds* in black livery at Nottingham Victoria, c1935. (T.G. Hepburn/Rail Archive Stephenson)

5438 *Worsley Taylor* at Grantham, c1936. (T.G. Hepburn/Rail Archive Stephenson)

62656 *Sir Clement Royds* at Manchester Central alongside a B1 which had taken over most of the Manchester–London trains via the former Great Central route after the Second World War, 1952. 62656 is in BR mixed traffic lined livery. (MLS Collection)

Operation

After being run-in from Gorton, the whole class, bar 438, was allocated to Neasden depot where they replaced a similar number of Atlantics. 438 remained at Gorton for a year and spent part of that outstabled at York, while one of the Neasden engines, 429, went to the GW shed at Oxford to enable crew familiarisation before working a royal train from Windsor to Lambton Castle in County Durham.

Deceleration of services because of the outbreak of the First World War did not take place immediately and the 'Directors' quickly made an impression with a number of fine runs being recorded and published in Cecil J. Allen's regular articles in the *Railway Magazine.* By 1914, 438 had joined its fellow members at Neasden.

62650 *Prince Henry* in BR plain black livery at Manchester Central with the 12.35pm stopping train to Northwich, 4 April 1953. (B.K.B. Green/MLS Collection)

62652 *Edwin A. Beazley* in BR lined mixed traffic livery on the turntable at Chester Northgate shed, 4 April 1953. (B.K.B. Green/MLS Collection)

The last survivor of the class, 62653 *Sir Edward Fraser,* at Northwich, 28 May 1955.
(N. Fields/MLS Collection)

		Marylebone–Leicester (GC period)				
		438 *Worsley Taylor*		438 *Worsley Taylor*		
		221/230 tons		303/330 tons		
		3.20pm Marylebone				
Miles	Location	Times	Speed	Times	Speed	Gradients
0	Marylebone	00.00		00.00		
5.1	Neasden	08.40	67	09.45	64½	1/100 R, 1/90 F
9.2	Harrow-on-the-Hill	12.50	47½	14.15	45	1/91 R
11.4	Pinner	15.10		16.40	61	1/176 F
13.7	Northwood	17.35	68	19.15	57½	1/145 R, 1/176 F
17.2	Rickmansworth	21.05	30*	22.50	35*	
19.4	Chorley Wood	24.40	39½	26.45	32	1/106 R
23.6	Amersham	31.00	40	34.45	32	1/105 R
28.8	Great Missenden	36.05		40.10	71½	1/160 F, 1/125 R
33.3	Wendover	40.35	76½	44.35	74	1/117 F
38	Aylesbury	44.40	60*	48.25	75	1/117 F
44	Quainton Road	51.05	50*	54.15	53*	
46.8	Grendon Underwood	53.45	65	57.15	62½	L, 1/176 F
48.8	Calvert	55.40	62	59.15	60	1/176 R
54.5	Finmere	61.20	57	65.15	50	1/176 R
59.3	Brackley	66.20	65/52	70.20	64½	1/176 F, 1/176 R
62.5	Helmdon	70.00	53	74.00	51	1/176 R
66.1	Culworth	73.35	70 ½	-		L, 1/176 F
67.2	Culworth Junction	-		78.50	71½	1/176 F
69.1	Woodford	76.15	65	80.20	66	L
71.5	Charwelton	78.35	60	82.45	58½	1/176 R
78.5	Braunston	84.30	80½	88.40	82	1/176 F
83.2	Rugby	88.30	64	92.35	63½	1/176 R
90	Lutterworth	95.20	71½/50½	99.10	64	1/176 F, 1/176 R
93.9	Ashby Magna	99.10	easy	103.05	70½	1/176 F
98.4	Whetstone	102.55	77½	106.40	82	1/176 F
103.1	Leicester	107.40	8¼ E	110.55	5 E	

		Nottingham - Leicester– Marylebone (GC period)						
		431 *Edwin A. Beazley*		434 *The Earl of Kerry*		430 *Purdon Viccars*		
		181/190 tons		215/230 tons		215/230 tons		
Miles	Location	Times	Speed	Times	Speed	Times	Speed	Gradients
0	Nottingham	00.00						
0.9	Arkwright Street	02.15						
4.4	Ruddington	06.25	65					L
11	Barnston	13.25	52					1/176 R
13.8	Loughborough	16.15	70					1/176 F
18.6	Rothley	20.50	62					1/264 R
21.3	Belgrave	23.25	60					1/176 R
23.4	Leicester	26.00	T					
		00.00	T	00.00		00.00		
4.7	Whetstone	07.20		06.50	53½	06.55		1/176 R
9.2	Ashby Magna	13.30	43½	12.35	47½	12.40	46½	1/176 R
13.1	Lutterworth	18.35	76½	17.10	66	17.25	71½	1/176 F
19.9.	Rugby	24.45		23.45	55	23.40	62	
24.6	Braunston	28.45	75	28.25	66	28.00	67	1/176 F, L
31.6	Charwelton	36.20	49½	36.40	50	36.15	47	1/176 R
34	Woodford	38.50	65	39.10		38.55		1/176 F
37	Culworth	41.30	70	41.55	69	41.50	66	1/176 F, 1/176 R
40.6	Helmdon	44.50	62	45.35		45.35		
43.8	Brackley	47.30	76½	48.40	75	48.40		1/176 F
48.6	Finmere	51.40	68	53.20		53.00		1/176 R, 1/176 F
54.3	Calvert	56.10	79	58.20	74	57.40	77½	1/176 F
56.3	Grendon Underwood	57.50		60.05		59.25		
59.1	Quainton Road	60.30	34*	62.40	pws	61.50		
65.1	Aylesbury	67.30		69.40	53	67.45	69	
69.8	Wendover	73.40	38½	76.05	40 ½	73.15	42½	1/117 R
74.3	Great Missenden	79.05	68	81.40	70½	78.40	74	1/125 F
79.5	Amersham	84.35	55	86.45	54	83.35	56½	1/160 R
83.7	Chorley Wood	88.30	70½	-	76½	-	76½	1/105 F
85.9	Rickmansworth	90.45	23*	92.40		89.25		
89.4	Northwood	95.40		97.00	52½	93.45	52½	1/176 R
91.7	Pinner	98.05		-	71½	-	67	1/145 R
93.9	Harrow-on-the-Hill	100.45	22*	101.30		98.20		
98	Neasden	104.55		105.30	70½	102.30	70½	1/91 F
103.1	Marylebone	111.45	4¼ E	111.35 (110½ net)	4½ E	109.05	7 E	

The Great Central 4-4-0s • 99

Leicester–Nottingham–Sheffield (GC period)

438 *Worsley Taylor*

221/230 tons

3.20pm Marylebone

Miles	Location	Times	Speed	Gradients
0	Leicester	00.00	T	
2.3	Belgrave	04.15		1/176 R
5	Rothley	07.15		1/176 F
9.8	Loughborough	11.35	76½	1/176 F
12.6	Barnston	14.20	60	1/176 R
19	Ruddington	19.35	75	1/176 F, L
22.5	Arkwright Street	23.00	sigs	
23.4	Nottingham	27.05	T	
0		00.00		
1.6	New Basford	04.15	T	1/130 R
	Bulwell Common	-	53½	1/330 F
5.8	Hucknall	09.45	46	1/130 R
10.8	Kirkby South Junction	16.05	48	1/132 R
17.9	Pilsley	23.40	65/49½	1/185, 1/132 F, 1/132 R
20.3	Heath	26.15		1/300 F
26.2	Staveley	31.50	70½	1/100 F
30	Killamarsh	36.00	*	colliery subsidence*
33.2	Woodhouse	39.35	60	1/185 F
36.2	Darnall	43.40		
38.2	Sheffield Victoria	46.20		1 ¾ E

It is to be noted that until the 1930s most logs were only recorded to the nearest five seconds. In a *Railway Magazine* article of early 1915 Cecil J. Allen described a run by the well-known Neasden driver Bailey on the 2.40am newspaper train which loaded to 180 tons and slipped a coach at Leicester. The august gentleman slipped by dating the run as June 1912, which must have been an error as the locomotive, 432 *Sir Edward Fraser*, was not built until October 1913. I'm assuming he meant June 1914. The run was conducted in pouring rain and a strong south-east gale. The train was through Rickmansworth ¾ minute early in 21¼ minutes, held a steady 39-40mph on the 1 in 105 climb through the Chilterns and with no rush down through Wendover, just 72mph, passed Aylesbury in 44 minutes 35 seconds, ½ minute early. Speed varied between 56½ on the long 1 in 176 through Brackley to Helmdon to the low 70s over the undulating grades before Woodford and a p-way slack at that location. Bailey then accelerated his engine hard down the 1 in 176 to Braunston, passed at just over 80mph, 64 and 72 before and after Rugby and another 80 – 82 in fact – at Whetstone, slipping its coach at Leicester in 106½ minutes from London against the 109 minutes schedule (104 minutes net for the 103.1 miles). After a severe p-way slack at Belgrave, 432 continued its merry way with 77½mph sustained before and after Loughborough, falling to exactly 60mph on the intervening 1 in 176 to Barnston Tunnel. Thus, Nottingham was passed still half a minute to the good in 130½ minutes (127 net) for the 127 miles. 50mph was held on the long 1 in 130/132 seven-mile climb to Kirkby Bentinck and with 75 through Staveley, a punctual arrival in Sheffield seemed assured until a three-minute signal stand was incurred at Woodhouse. Sheffield was reached in 175 minutes 40 seconds, 1¾ minutes late (169 minutes net for the 164.6 miles).

Another tightly timed train was the 4.55pm Marylebone, which was allowed just 46 minutes for the 38 miles to Aylesbury and 65 minutes for the 65 miles on to Leicester. 437 *Charles Stuart-Wortley* got its

165 tons to the Aylesbury stop in 44¾ minutes despite a p-way restriction just after Wendover. 67mph was achieved before Neasden, 72½ at Northwood, and on the climb to Amersham 437 sustained a steady 42-43mph. 74 before and 75mph after Great Missenden got time in hand for the planned track work. The mile-a-minute run from Aylesbury to Leicester was completed in exactly that – 64 minutes 50 seconds requiring speeds of 53-54mph on the long 1 in 176 gradients and 81 at Braunston and 77 at Whetstone. 433 *Walter Burgh Gair* on the same train clipped a minute and a half off that time, the improved time a result of a much faster climb on 1in 176 through Brackley to Helmdon at 57.6mph and an even faster descent to Braunston which must have been in the high 80s. Cecil J. Allen noted in his articles an 86, three 85s, three 83s, four 82s, a 79, four 78s and six 75s at different locations with 'Director' runs during that initial period..

Some expresses were routed via the GW/GC joint line via High Wycombe and Ashendon Junction. The 10am Marylebone was one of them and a couple of logs timed in the Great Central era are below:

High Wycombe–Woodford

		430 *Purdon Viccars* 175 tons Driver Bailey, Neasden		432 *Sir Edward Fraser* 135 tons		
Miles	Location	Times	Speed	Times	Speed	Gradients
0	High Wycombe	00.00		00.00		
2.2	West Wycombe	04.40		04.10		
5	Saunderton	08.05	50 ½	07.15	56½	1/164 R
8.1	Princes Risborough	11.20	75/82 ¾ E	10.20	77½ 1 ¾ E	1/88 F
13.5	Haddenham	15.25	75	14.45	68	1/176 F, L
17.5	Ashendon Jcn	18.30	80½	18.10	74	L
18.9	Wotton	19.40	75	19.25	68	
20.9	Akeman Street	21.10	82	21.10	64½	
23.4	Grendon Underwood	23.10	75/77 2¾	23.30	64/70	
25.4	Calvert	24.45	71/75	25.20		1/176 R/ 1/176 F
31.1	Finmere	29.45	62	30.35	58 ½	1/176 R
35.9	Brackley	34.00	72	35.05	72	1/176 F/ 1/176 R
39.1	Helmdon	37.05	62	38.20	57	1/176 R
42.7	Culworth	40.10	73	38.20	65	
44	Culworth Junction	41.10	78	44.00	sigs stand	1/176 F
45.7	Woodford	43.55	5 E	48.45 (45 ¼ net)	1 ¼ E	

Another run in the Up direction via the GC direct route by the redoubtable Bailey was on the 3.25pm from Sheffield. The run was timed from Leicester, with 430 *Purdon Viccars* and 210 tons. Leaving Leicester 430 climbed the 1 in 176 through Whetstone to Ashby Magna at 55mph, sped past Lutterworth at 82 and cleared Rugby (19.9 miles) in 21¼ minutes. After 75 before Braunston, 430 fell to 51 minimum at Charwelton and stopped at Woodford in 35minutes 25 seconds (34 miles). The 9.8 miles to Brackley took 11 minutes 40 seconds start-to-stop, by which time a late start had been recovered. The 59.3 miles to Marylebone then took 65½ minutes with 81mph at Grendon Underwood, easy running past Quainton Road and Aylesbury, 48mph on the 1 in 117 through Wendover, 72 at Great Missenden, 54 at Amersham summit and 75 before the Rickmansworth 25mph slack (35.7 miles, 37 minutes 45 seconds).

430 *Purdon Viccars* headed an excursion from Nottingham to London via High Wycombe just before the First World War, calling to pick up passengers at Lutterworth and Rugby. It had seven coaches, 185 tons, and ran the 6¾ miles from Lutterworth to Rugby in 8¾ minutes start-to-stop, with a maximum of 82mph at Shawell. After the Rugby stop 49mph was sustained on the six mile climb to Charwelton and after Helmdon the driver let 430 fly, just touching 90mph below Brackley and again after Finmere. Grendon Junction was passed in 34 minutes 55 seconds for the 36.2 miles from Rugby. Marylebone was reached in 91 minutes 40 seconds (88 net) for the 87¾ miles.

Finally, one log of a run over the Woodhead route: 435 *Sir Clement Royds* had 225 tons on the 8.20am Manchester London Road to Marylebone.

Manchester–Penistone
435 *Sir Clement Royds*
6 chs/225 tons

Miles	Location	Times	Speed	Gradients
0	Manchester London Rd	00.00	T	
5	Guide Bridge	10.05		1/100, 133 R
9.8	Mottram	19.15	43	1/97, 143 R
12.7	Hadfield	25.05	32½	1/100 R
17.7	Crowden	33.15	34	1/117 R
19.7	Woodhead	36.40	38½	1/117 R
22.2	Dunford Bridge	39.55		
27.2	Penistone	46.50	T	

An unknown member of the class drawing into Leicester station with a down express, c1914. (W.A. Brown/MLS Collection)

429 passing Abbey Lane Sidings near Leicester with the Great Western royal train, c1914.
(J. Bradshaw/MLS Collection)

5437 *Prince George* with a down Great Central route express on Charwelton troughs, c1928.
(F. Moore/MLS Collection)

The Great Central 4-4-0s • 103

5436 *Sir Berkeley Sheffield* at speed on the Great Central Main Line, c1928. (F. Moore/MLS Collection)

5437 *Prince George* leaving Sheffield Victoria with a Marylebone express, 14 September 1929. (G. Coltas/MLS Collection)

It soon became obvious that Robinson had succeeded in producing an outstanding engine suited to the express passenger requirements of the company and its long sustained gradients of 1 in 176. Speeds in the 80s were common on most runs and could even reach 90mph on the long downhill stretches such as the approach to Leicester at Whetstone. These ten locomotives remained at Neasden throughout the First World War, 434 being the first to be transferred elsewhere (to Gorton) in 1918. Robinson built eleven 'Improved Directors' (class 11F) between 1919 and 1922 and these were all allocated to Neasden displacing the 11E class, 434 and 435 going to Sheffield and 436 and 438 to Woodford. When all the 11Fs were delivered, 437 moved to Gorton and 429, 430 and 432 went to Annesley working from Mansfield and Nottingham to Leicester and to Sheffield.

After the Grouping the whole ten, now class D10, were based at Gorton, although in the following year, the D10s and D11s (the former 11Fs) were split between Gorton and Neasden with, surprisingly, six D10s now at Neasden (5429–5434). Further exchanges of members of the two classes between Gorton and Neasden continued, though the crews seemed to prefer the D11s – possibly because of their more 'luxurious' cabs! D11s therefore seemed to be allocated to the fastest and heaviest trains although the D10s were perfectly competent understudies. 'Directors' appeared occasionally at King's Cross on excursion work or on the short-lived Sheffield Pullman, with fill-in trips on stopping trains to Peterborough or the Cambridge line.

A couple, 5431 and 5435, moved to Sheffield in the early 1930s and 5431 went on the Leeds Copley Hill working an occasional Pullman train but mainly Leeds expresses to Doncaster before handing over to a Gresley Pacific. 5432 and 5434 joined 5431, remaining at Leeds until 1938. The allocation of the ten engines in 1934 was:

Gorton:	5429, 5433, 5438
Neasden:	5430, 5436, 5437
Sheffield:	5431, 5435
Leeds (Copley Hill):	5432, 5434

The D10s, along with the D11s and C4 Atlantics were still in command of the main former GC services through the 1920s and early 1930s until the Gresley B17s were introduced to the line. The schedules had been accelerated to 109 minutes for the Marylebone–Leicester 103.1 miles and as loads increased from the six or seven coach loads of the GC era, these engines were worked hard to keep the schedules. I table below three logs from Cecil J. Allen's *Railway Magazine* articles, the runs believed to have occurred around 1930, although CJA did not date them.

Marylebone–Leicester Central, c1930

		4.55pm M'bone			4.55pm M'bone			3.20pm M'bone			
		5431 *Edwin A. Beazley*			5431 *Edwin A. Beazley*			5430 *Purdon Viccars*			
		7 chs, 175/180 tons			175/180 tons			239/250 tons			
Miles	Location	Times	Speed		Times	Speed		Times	Speed		Gradients
0	Marylebone	00.00			00.00			00.00			1/100 R
5.1	Neasden	08.25	pws	T	08.50	pws	¼ L	09.35	62½	½ L	1/90 F
9.2	Harrow-on-the-Hill	14.55		2 L	14.30		1½ L	14.35	39½	½ L	1/91 R
11.4	Pinner	17.20			17.10			18.25	pws		1/176 F
13.7	Northwood	19.35			20.05			21.50	66		1/145 R, 1/176F
17.2	Rickmansworth	23.15	*	1¼ L	23.40	*	1¾ L	25.35	*	2½ L	
19.4	Chorley Wood	26.55	41		27.00	41		29.10	37 ½		1/106 R
23.6	Amersham	30.00	42½		30.10	41½		32.40	39		1/105 R

		Marylebone–Leicester Central, c1930									
		4.55pm M'bone			4.55pm M'bone			3.20pm M'bone			
		5431 *Edwin A. Beazley*			5431 *Edwin A. Beazley*			5430 *Purdon Viccars*			
		7 chs, 175/180 tons			175/180 tons			239/250 tons			
Miles	**Location**	**Times**	**Speed**		**Times**	**Speed**		**Times**	**Speed**		**Gradients**
28.8	Great Missenden	37.50		2¾ L	38.15		3¼ L	41.20	70/50½	5¼ L	
33.3	Wendover	42.20	80½		43.15	67		45.45	76½		1/117 F
38	Aylesbury	46.05		1 L	47.30		1½ L	49.40		3¾ L	1/117 F
44	Quainton Road	52.15		¼ L	53.35		1½ L	55.50		2¾ L	
46.8	Grendon Underwood	55.05			56.10	69		58.50	62½		L, 1/176 F
48.8	Calvert	57.00			57.55	64½		60.45			1/176 R
54.5	Finmere	62.45	52		63.10	55½		66.55	48		1/176 R
59.3	Brackley	67.40			67.50			72.20	64½		1/176 F
62.5	Helmdon	71.15	53		71.05	57		75.55	51		1/176 R
66.1	Culworth	74.40			74.30			79.30	69		L, 1/176 F
69.1	Woodford	77.15		1¼ L	77.10		1¼ L	82.15		4¼ L	L
71.5	Charwelton	79.40			79.40			84.45			1/176 R
78.5	Braunston	85.25	85		85.50	80½		91.10	75		1/176 F
83.2	Rugby	89.20	65		89.55	62½		95.30	56½		1/176 R
90	Lutterworth	95.55			96.26	70/54		102.30			1/176 F, 1/176R
93.9	Ashby Magna	99.45			100.25			106.15			1/176 F
98.4	Whetstone	103.15	85		104.10	80½		109.50	82		1/176 F
103.1	Leicester	107.30		¼ E	108.35		½ L	114.05		5 L	

In the first log of 5431, the only speeds that were noted were on the climb and descent from Amersham, and the speed at Helmdon summit. The rest I have estimated from calculating the average speeds. There is no doubt that high speeds of at least 85mph were achieved at both Braunston and Whetstone, possibly higher. The schedule from Quainton Road to Woodford (25.1 miles, 25 minutes) was tough as it included long climbs of 1 in 176 and only the second run of 5431 maintained this schedule, 5430 with 250 tons losing nearly two minutes on this section. 5431 featured in one of the fastest Leicester–Nottingham runs in November 1936 when, with 195 tons, it ran the 26.4 miles in 24 minutes 45 seconds start-to-stop, averaging 82.2mph over the 5.6 miles between East Leake and Ruddington on the 1 in 176 falling from Barnston Tunnel. Since it had only one mile to recover from three miles of 1 in 176 up from Loughborough, the passing speed at East Leake is not likely to have been much over 70mph, so the high 80s, if not 90mph is probable between Gothan Junction and Ruddington.

By the 1930s, when six of the class were based at Gorton, the D10s were more common on the Woodhead route with Manchester–Sheffield trains than on the through London workings. A couple of typical logs are shown below, not spectacular but showing the timings and speeds necessary to maintain the 59 minute schedule for the 41.3 heavily graded miles.

		Sheffield Victoria–Manchester London Road					
		5438 *Worsley Taylor*		5433 *Walter Burgh Gair*			
		175/180 tons		175/180 tons			
Miles	Location	Times	Speed	Times	Speed		Gradients
0	Sheffield Victoria	00.00		00.00			
1.2	Neepsend	03.05		03.20			1/132 R
2.9	Wadsley Bridge	05.25	43½	06.05	37		1/132 R
4.9	Oughty Bridge	08.30	46	08.55	42		1/132 R
7.9	Deepcar	12.40	42	13.40	38		1/120 R
8.8	Wortley	13.55	46	15.10	36		1/131 R
12.9	Penistone	19.30	44 ½ E	21.10	41	1¼ L	1/160R, 1/100 R
16.7	Hazlehead Bridge	24.50	42½	26.30	43		1/124 R
18.9	Dunford Bridge	28.05	40½ 1 E	29.55	37½		1/135 R
22.1	Woodhead	31.45	1¼ E	33.50		¾ L	1/201 F
24.1	Crowden	34.00	50 easy	35.50	65 easy		1/190 F
29.4	Dinting	40.55		41.05			1/100 F
31.5	Mottram	43.35		43.50			1/127 F
33	Godley Junction	45.35		46.15			1/145 F
36.3	Guide Bridge	49.40	¼ E	50.95		T	1/97 F
41.3	Manchester L. Road	57.20	1¾ E	57.10	1¾ E		

The locomotive performance interest is purely on the climb to Woodhead Tunnel in both directions as speed is restrained down the curvature from the moors to the Manchester and Sheffield suburbs. Both runs were unchecked into Manchester and therefore did not need the couple of minutes recovery time approaching the terminus. Later in the 1930s loads had grown and 5438 was faced with 10 coaches out of Sheffield, reduced to 7 after the Halifax potion was detached at Penistone.

A major redistribution of former GC engines took place in 1936/7 when an allocation of B17 'Footballers' was made to Neasden and Gorton. The Gorton D10s and 5430 and 5436 from Neasden joined the pair at Sheffield, in turn displacing the Robinson D9s and taking up their duties to York, Hull, Lincoln and Nottingham, as well as occasional sallies south to Leicester or Banbury. Neasden's remaining D10, 5437, and 5431 from Sheffield were transferred to Annesley in 1937, also to replace D9s, one being used as standby engine at Nottingham Victoria to replace or assist any locomotive in trouble. Retford received 5438 for a regular turn to Cleethorpes returning on the heavy evening mail train.

During the Second World War, freight work increased significantly and the D10s found themselves involved working as far north as Newcastle on goods trains. Three – 5430, 5432 and 5438 – were initially sent to Doncaster, then 5430 and 5438 worked from Langwith on military traffic from the Royal Ordnance Factory at Ranskill. From August 1943 the whole class was based at Sheffield from where they could be used on a whole variety of both passenger and freight work. Because of the huge increase in train weights, undoubtedly they would have been useful for piloting assistance. Their diagrams would have included stopping trains in the Doncaster, Rotherham and Mexborough area, many of which could consist of

Sheffield Victoria–Guide Bridge

5438 Worsley Taylor

10 chs, 339/360 tons to Penistone

7 chs, 241/260 tons from Penistone

Miles	Location	Times	Speed	Gradients
0	Sheffield Victoria	00.00	T	
1.2	Neepsend	03.05		1/132 R
2.9	Wadsley Bridge	06.25	30	1/132 R
4.8	Oughty Bridge	10.10	32	1/132 R
7.9	Deepcar	16.15	30	1/120 R
8.8	Wortley	18.00	31/35	1/131 R
12.9	Penistone	25.55	T	
0		00.00		
3.6	Hazlehead Bridge	08.05	38	1/100 R, 1/130 R
6	Dunford Bridge	11.40	40/ pws 20*	1/135 R
9.2	Woodhead	16.10		1/185 F
	Crowden	-	66 sigs, severe	1/117 F
23.4	Guide Bridge	38.25		

initial sojourn was temporary, by the end of 1947 the allocation was:

Liverpool
  Brunswick: 2653, 2656, 2658
Trafford Park: 2651
Northwich: 2650, 2652, 2655
Sheffield: 2654, 2657, 2659

5436 had 215 tons on a Manchester–Liverpool train in 1947, reaching Liverpool (34 miles) in 42 minutes net with stops at Warrington and Farnworth with unusually high maxima of 67mph at Springfield and Mersey Road and 60mph minimum at Hunts Cross on the final Farnworth–Liverpool section arriving a minute early. 2658 was working the 12.30pm Liverpool–Manchester with 8 coaches later in 1947, climbed the 1in 195 to Hunts Cross at 35mph, with 54 down the 1 in 185 to Hough Green, dropping nearly 2 minutes on the 18 minute schedule to Farnworth (13.4 miles).

quite substantial rakes of coaches. Ten coach trains were timed by Mr J. Wedgwood during the late 1930s and the war years with 31 minutes schedules for the 18½ miles with stops at Mexborough and Rotherham pre-war and 38 minutes from the early 1940s. 5435 took 30¼ minutes on 22 May 1937, 5437 30½ minutes on 23 July 1938 5433, just under 36 minutes (33½ net) on 31 July 1943 and 34¾ net on 16 June 1945, a run with severe signal checks around Rotherham.

After the war, the first D10s were sent to the CLC system, once again taking over from the D9s. 5435 was the first, sent to Liverpool Brunswick in May 1946, the renumbered 2653 (ex-5432) going to Trafford Park at the end of the year. Although their

5434 *The Earl of Kerry* leaving Penistone with 12.15pm Marylebone–Manchester, 25 August 1932. (MLS Collection)

5430 *Purdon Viccars* with a Sheffield–Marylebone express near Parkend Colliery, South Yorkshire, c1932. (R. Carpenter/MLS Collection)

5438 *Worsley Taylor* at Hyde Junction with the 3.50pm Manchester–Marylebone, 2 June 1934. (MLS Collection)

The 6.2 miles on to Warrington was better, completed in 9½ minutes, recovering half a minute (max 56mph at Sankey) and the rest of the lost time was recouped on the run to Manchester despite a 40mph p-way restriction at Risley Moss, with a maximum speed of 54mph at Flixton.

2656 soon moved from Brunswick to Trafford Park. This meant that the four at Liverpool and Manchester shared some of the express work as well as the stopping trains between the two cities, as well as sometimes working through to Hull. Their condition recovered around this time, with some, including Brunswick's 62658, being out-shopped in the new BR mixed traffic livery. By October 1949 all ten were active on the CLC routes. At the end of 1949 the allocation was:

Liverpool
  Brunswick: 62651, 62653, 62656-62658
Trafford Park; 62654
Northwich: 62650, 62652, 62655

This was hardly ideal work for the 4-4-0s and their performances were fairly humdrum as the route was not really suitable for high speed running. To give an example of their post-war use, I show three logs below from the Railway Performance Society's archives – all were timed by Gerald Aston. 62657 was clearly overloaded on this schedule, 62656 looks as though it was struggling and the signallers didn't give 62653 much of a chance.

Liverpool Central–Manchester Central

Miles	Location	62657 *Sir Berkeley Sheffield* 10 chs, 332 tons 4.30pm Liverpool – Hull 12.8.1950			62656 *Sir Clement Royds* 7 coaches, 210 tons 6.30pm L'pool – M'chester 17.12.1951			62653 *Sir Edward Fraser* 6 chs, 162 tons 4.20pm L'pool – M'chester 16.7.1953			Gradients
		Times	Speed		Times	Speed		Times	Speed		
0	Liverpool Cen	00.00			00.00			00.00			
2.6	St Michael's	05.51	36		05.20	39		05.33	42		1/260 R
4	Mersey Road	08.25	46		08.18	37		07.12	53		1/400 R
4.9	Cressington	09.24	44	1½ L	09.19	45/42	1¼ L	08.00	56	T	
7.1	Hunt's Cross	12.44	39	1½ L	13.04		1½ L	11.24	sigs 30*/35		1/195 R
					00.00		1½ L				
8.3	Halewood	14.29	62½	2½ L	03.06	53		13.01	62½	1 L	1/185 F
12.2	Farnworth	19.06		2½ L	08.15			17.26		1 L	
		00.00		2¾ L	00.00		2¾ L	00.00		1¾ L	
1.8	Widnes E Jcn	03.41	50½	2½ L	03.36	48	2¾ L	03.20	52		1/230 F
	Sankey	05.30	64½		05.31	58		05.11	60		1/158 F
5	Sankey Jcn	06.57	28*	2½ L	07.03	31*	2¾ L	06.33	34*	1¼ L	
6.1	Warrington	10.03		3¾ L	09.46		3½ L	08.52		½ L	
		00.00		4 L	00.00		2½ L	00.00		¾ L	
1.5	Padgate Jcn	03.34	41/ pws 15*	4½ L	03.19	44	2¾ L	03.12		1 L	
								04.02/04.32	sigs		

Liverpool Central–Manchester Central

Miles	Location	62657 *Sir Berkeley Sheffield* 10 chs, 332 tons 4.30pm Liverpool – Hull 12.8.1950			62656 *Sir Clement Royds* 7 coaches, 210 tons 6.30pm L'pool – M'chester 17.12.1951			62653 *Sir Edward Fraser* 6 chs, 162 tons 4.20pm L'pool – M'chester 16.7.1953			Gradients
		Times	Speed		Times	Speed		Times	Speed		
6.1	Glazebrook	09.38	pws 30*		08.33	46½ / 44		11.00	47		1/240 R
	G'brook E Jn	10.54	49	6 ½ L	09.46	49	3 ½ L	12.12	50	4¼ L	L
	Irlam	12.00	50		10.54	50		13.25	56		L
8.2	MP 26½	-	45		-	41		-	40		1/135 R
	Flixton	14.31	61		13.41	56		16.20	55		1/135 R
10.7	Urmston	-	30* sigs		-	pws 15*		19.44		3¾ L	L
								00.00		4 L	
12.5	Trafford Park	19.45	39	8¼ L	19.43		5¼ L	02.30	40		
	Throstle N E Jn	20.43	46		20.40	45/pws 10*		24.32	42		
15.7	Manchester	24.13		8½ L	24.23		5 ½ L	27.30		2½ L	

2655 *The Earl of Kerry* at Ashley with a late afternoon Manchester Central–Chester stopping train, 20 March 1948. The condition of the former GC 4-4-0s working the CLC local services in the post-war years is illustrated all too clearly in this photo.
(J.D. Darby/MLS Collection)

After nationalisation the CLC lines became part of the London Midland Region and some LMS 4-4-0s of both 2P and 4P classes were drafted in as replacements but were found inadequate and GC D11s and ex GE D16s were sent to operate the services between Liverpool and Manchester. The D10s were then restricted to the Manchester-Chester route, most working from Northwich shed from which eight were withdrawn between 1953 and 1955. 62654 and 62657 were both withdrawn from Trafford Park in 1953. The final allocations in the 1950s were:

	1950	1952	1954
62650	Northwich	Northwich	Northwich
62651	Trafford Park	Northwich	
62652	Northwich	Northwich	Northwich
62653	Trafford Park	Trafford Park	Northwich
62654	Trafford Park	Trafford Park	
62655	Northwich	Northwich	
62656	Trafford Park	Trafford Park	Northwich
62657	Trafford Park	Trafford Park	
62658	Trafford Park	Trafford Park	Northwich
62659	Trafford Park	Trafford Park	Northwich

Its new BR number just about visible, 62655 *The Earl of Kerry* with a Manchester–Chester stopping train at Ashley, 18 April 1949.
(G.M. Shoults/MLS Collection)

62655 *The Earl of Kerry* and 62651 *Purdon Viccars* at Manchester Central with race specials for Aintree, 1952.
(N. Harrop/MLS Collection)

62659 *Worsley Taylor* is unusually coupled with Ivatt 2MT 2-6-2T 41235 as it departs from Knutsford with a Manchester–Chester train, 5 June 1954. LMR motive power was taking over around this time on the CLC lines as the last GC 4-4-0s were being withdrawn.
(B.K.B. Green/MLS Collection)

The D11/1, J.G. Robinson, 1919

An order for five further 'Directors' was made in 1916 and cancelled in favour of a new class of 4-6-0s, the 'Lord Faringdon' class, of which the first example was built at Gorton in 1917. However, no more were built until 1920 and the 'Director' 4-4-0 order was remade in 1919, possibly as a result of the 4-6-0, though better than the 'Sir Sam Fay' class, still not meeting expectations. Also in the post-war period, train loadings had fallen within the compass of 4-4-0 haulage. Five new engines known as 'Improved Directors' and classified '11F' were constructed at Gorton in 1919 and 1920 and numbered 506–510. Six more, 501–505 and 511, were built at Gorton in 1922 just before the Grouping. The main improvements over the '11Es' were the provision of inside admission for the piston valves and cabs with side windows. Their dimensions were:

Cylinders (2 inside):	20 x 26in
Coupled wheel diameter:	6ft 9in
Bogie wheel diameter:	3ft 6in

Stephenson valve gear with 10" piston valves

Boiler pressure:	180lbs psi
Heating surface:	1,543sq ft (incl superheater 209sq ft)
Grate area:	26.5sq ft
Axleload:	19 tons 18 cwtr
Weight (Engine):	61 tons 3 cwt
(Tender):	48 tons 6 cwt
(Total):	109 tons 9 cwt
Water capacity:	4,000 gallons
Coal capacity:	6 tons
Tractive effort:	19,644lbs

They were named:

506	*Butler-Henderson*
507	*Gerard Powys Dewhurst*
508	*Prince of Wales*
509	*Prince Albert*
510	*Princess Mary*
501	*Mons*
502	*Zeebrugge*
503	*Somme*
504	*Jutland*
505	*Ypres*
511	*Marne*

506 *Butler-Henderson*, the first '11F Improved Director' as constructed at Gorton in 1919. (MLS Collection)

510 *Princess Mary* shortly after construction at Gorton in May 1920. (Bob Miller/MLS Collections)

504 *Jutland* at Neasden, in Great Central livery but with the L&NER symbol on the tender, 1923. (MLS Collection)

5502 *Zeebrugge* at Neasden in LNER apple green livery, at the end of 1924.
(F. Moore/MLS Collection)

5501 *Mons* at Neasden after valance removal, renumbering and painting in the LNER apple green livery, c1925.
(F. Moore/MLS Collection)

They were, despite the imminence of the Grouping, brought out in Great Central livery and numbers, although the penultimate three, 503–505, had L&NER inscribed on the tender and 511 was painted in LNER lined apple green. They were renumbered 5501–5511 in 1924 and like other LNER 4-4-0s, painted black in 1928. After exposure to cylinder long travel valve events at the GWR Castle/LNER Pacific exchange in 1925, Gresley eventually became persuaded of the development and fitted 5505 with new cylinders and long travel valves in January 1937, an improvement already introduced earlier for his Pacifics and larger locomotives like the former Great Eastern Holden 4-6-0s (B12/3). Tests revealed a 5 per cent economy in coal consumption which justified further conversions as new cylinders were required.

Eventually, all except 5506 were converted between 1944 (5507) and as late as 1952 (62665). They were prone to cracked frames and most were strengthened at the back end in the early 1930s, with one, 62663, receiving this major repair as late as 1956, only four years before withdrawal. Their Robinson chimneys were replaced by shorter Gresley 'flowerpots' and later in the 1930s by cast chimneys similar to the original Robinson ones. They were renumbered 2660–2670 in the LNER 1946 renumbering scheme and 62660–62670 after 1948. The first of the GC built locomotives was withdrawn in May 1959 (62665) and the last, 62666, in December 1960.

Operations

In many ways, the initial use of the 11Fs followed that of the 11Es – the first five were allocated to Neasden, followed by the rest of the class after their construction in 1922. They shared the London–Leicester–Nottingham–Sheffield work with the Leicester based Atlantics until 1924 when a switch of D10s and D11s between Neasden and Gorton took place, with the now Gorton based 5506–5511 sharing work over the Woodhead route to Leicester, Retford and Sheffield with the B2 ('Sir Sam Fay') 4-6-0s, as well as the London work from the Manchester end.

Some early runs with the new D11s were recorded in the last couple of years before the Grouping and immediately afterwards. A sample is given below of three runs on the 3.20pm Marylebone and another three on the 2.32am newspaper train, a lighter load but tightly timed and noted for some of the highest speeds recorded by the 'Directors'.

5501 *Mons* in LNER black livery, c1930.
(W. Potter/MLS Collection)

5502 *Zeebrugge* at Brunswick shed, Liverpool, July 1937. (W. Potter/MLS Collection)

5505 *Ypres* in run-down immediate post-war condition at Sheffield Darnall shed, 14 October 1945. (MLS Collection)

2663 *Prince Albert* after the LNER 1946 renumbering, at Immingham shed, 8 June 1947.
(N. Fields/MLS Collection)

62667 *Somme* in ex-works condition and repainted in the BR mixed traffic lined black livery at Sheffield Midland, 7 June 1953.
(J.D. Darby/MLS Collection)

The Great Central 4-4-0s • 119

A year later 62667's condition has significantly deteriorated. It is pictured at Lincoln St. Marks shed, July 1954.
(T.K. Widd/MLS Collection)

62669 *Ypres* in BR plain black livery at Manchester London Road, 5 March 1959.
(B.K.B. Green/MLS Collection)

		Marylebone–Leicester Central, c1922/3						
		3.20pm M'bone 510 *Princess Mary* 296/320 tons		3.20pm M'bone 507 *Gerard Powys Dewhurst* 220/230 tons		3.30pm M'bone 504 *Jutland* 279/295 tons		
Miles	Location	Times	Speed	Times	Speed	Times	Speed	Gradients
0	Marylebone	00.00		00.00		00.00		1/100 R
5.1	Neasden	09.40	57 ½	08.50	65	09.15	66½	1/90 F
9.2	Harrow-on-the-Hill	14.40	40	13.15	43½	13.22	50	1/91 R
11.4	Pinner	17.05	59	15.35		15.35	65½	1/176 F
13.7	Northwood	19.50	50/62½	18.05	52/68	17.55	56/70	1/145 R, 1/176F
17.2	Rickmansworth	23.30	*	21.40	*	21.28	45	
19.4	Chorley Wood	27.10	32½	25.15	39½	24.34	42	1/106 R
23.6	Amersham	34.50		31.35	40	32.05	31	1/105 R
28.8	Great Missenden	40.20	69	36.40		37.24	74	
33.3	Wendover	45.10	79	41.10	74	41.46	74/82	1/117 F
38	Aylesbury	49.05		45.15	55	45.20		1/117 F
44	Quainton Road	55.00	50*	51.50	50*	50.48	55*	
46.8	Grendon Underwood	57.45		54.35	61	53.43	66	L, 1/176 F
48.8	Calvert	59.45	62½	56.35		55.33	65½	1/176 R
54.5	Finmere	66.00		62.40		60.52	56	1/176 R
59.3	Brackley	71.40	60/43	67.55	62/53	65.35	69	1/176 F
62.5	Helmdon	75.50	45	71.45	47½	68.35	56	1/176 R
66.1	Culworth	79.40	68	75.30	69	-	75	L, 1/176 F
69.1	Woodford	82.36		78.15		74.52	68½	L
71.5	Charwelton	85.05	55	80.40		77.00	64	1/176 R
78.5	Braunston	91.15	80½	86.40	82	82.39	84	1/176 F
83.2	Rugby	95.10	64½	90.40	65	86.26	65½	1/176 R
90	Lutterworth	101.45	72/54	97.10	72/53½	92.55	62	1/176 F, 1/176R
93.9	Ashby Magna	105.40		101.00		96.50	69½	1/176 F
98.4	Whetstone	109.15		104.30	79	100.25	76½	1/176 F
103.1	Leicester	114.00	T	109.05	T	104.52	4 E	

Princess Mary was still on the wartime 114 minute schedule with a substantial load for a 4-4-0, whereas *Gerard Powys Dewhurst* and *Jutland* were operating to the post-war restored 109 minute schedule. *Jutland's* run was exceptional with nearly 300 tons.

		Marylebone–Leicester Central, Newspaper Train, c1922/3						
		2.32am M'bone 509 *Prince Albert* 140/155 tons		2.32am M'bone 510 *Princess Mary* 185/200 tons		2.32am M'bone 501 *Mons* 160/175 tons		
Miles	Location	Times	Speed	Times	Speed	Times	Speed	Gradients
0	Marylebone	00.00		00.00		00.00		1/100 R
5.1	Neasden	08.25	69	08.20	71	08.25	68½	1/90 F
9.2	Harrow-on-the-Hill	12.35	45	12.25	44	12.35	45	1/91 R
13.7	Northwood	17.35	56	17.20	51/73 ½	17.25	53/69	1/145 R, 1/176 F
17.2	Rickmansworth	21.25	*	20.40	*	20.50	*	
19.4	Chorley Wood	24.35		23.40	46	24.35		1/106 R
21.6	Chalfont	27.50	40	26.45	43	28.10	34	1/105 R
23.6	Amersham	30.35	43½	29.20	46	31.35	38½	1/105 R
28.8	Great Missenden	35.25	78	34.10	78	36.55	70½	
33.3	Wendover	39.50	80	38.30	85	41.15	80½	1/117 F
38	Aylesbury	43.50	¼ E	42.20	60* 1¾ E	45.05	51* 1 L	1/117 F
44	Quainton Road	50.05	2 E	48.40	3¼ E	52.05	T	
46.8	Grendon Underwood	52.25	67	51.00	67 ½	54.45	65	L, 1/176 F
48.8	Calvert	54.35	50/70	52.55	54/69	56.45	53½/65	1/176 R, 1/176 F
54.5	Finmere	60.25	50	58.45	51	63.20	47	1/176 R
59.3	Brackley	65.55	1 E	64.15	2¾ E	68.40	1¾ L	1/176 F
		00.00		00.00		00.00		
3.2	Helmdon	05.05	52	05.25	50	05.20	50	1/176 R
6.8	Culworth	08.35	72	08.50	73½	09.00	71½	L, 1/176 F
9.8	Woodford	11.30		11.20		11.40		L
12.2	Charwelton	13.40	60	13.35	60	13.55	61	1/176 R
19.2	Braunston	19.45	85	19.40	80	19.35	90	1/176 F
23.9	Rugby	24.00	T	24.05	T	23.40	1½ L	1/176 R
		00.00		00.00		00.00		
6.8	Lutterworth	08.10		08.10	sigs 25*	08.10		1/176 F, 1/176R
10.7	Ashby Magna	11.50		13.20		11.35		1/176 F
15.2	Whetstone	15.25	75	17.25	71	14.55	82	1/176 F
19.9	Leicester	19.55	T	22.35	2½ L	18.55	½ L	

502 *Zeebrugge* at Hazlehead Bridge with a Manchester–Marylebone express, 1922.
(MLS Collection)

505 *Ypres* was timed on the 4.55pm Marylebone from Aylesbury around the same time and ran the 65.1 miles to Leicester in 63 minutes 25 seconds with 165 tons in tow. Highlights were the maintenance of 58mph on the long 1 in 176 climb to Helmdon, 78 at Culworth Junction and just over 80 at Braunston before 505 was eased to avoid too early an arrival. 507 *Gerard Powys Dewhurst* with 230 tons completed the 23.4 miles from Leicester to Nottingham in 24 minutes 50 seconds with 77½mph at Loughborough, 60 minimum at Barnston Box and another 75 at Ruddington, arriving a few seconds early. 5509 *Prince Albert*, around 1924 in the opposite and harder direction, got its 225 tons to Leicester in 25 minutes 39 seconds just beating the 26-minute schedule in that direction, after a slow start averaging 59.2mph on the four-mile climb to Barnston Box and averaging 83mph from East Leake to Loughborough and 88 from Loughborough to Rothley which suggests the speed in the dip at Loughborough must have been at least 90mph.

In 1927, D11 5511 was tested on the East Coast Edinburgh Pullman (named *The Queen of Scots* from 1928) between Leeds and King's Cross following the less than successful use of GC B3 4-6-0s on that service and after a promising trial operation, 5507 joined 5511 at Copley Hill (Leeds), sharing the Pullman work with ex-GNR C1 Atlantics. Between 1928 and 1932 5501–5503, 5506, 5507, 5510 and 5511 all had spells at Copley Hill, 5506 and 5507 spending most of those years there. Cecil J. Allen described an early couple of runs on the Up *Harrogate Pullman* in one of the 1927 *Railway Magazines*:

Miles	Location	Harrogate–King's Cross Pullman				5511 *Marne*			Gradients
		5511 *Marne*				7 Pullmans, 282/290 tons			
		6 Pullmans, 248/255 tons							
		Times	Speed			Times	Speed		
0	Harrogate	00.00		T		00.00		T	1/86 F
5	Spofforth	06.50	70½ /55			-			1/116 R
8	Wetherby Junction	09.40	69		1¾ E	-			1/171 F
14.4	Tadcaster	15.25	40*			16.10			
19.1	Church Fenton	21.20	35*		3¾ E	22.50		2¼ E	L
25	Burton Salmon	28.40				30.45			L
28.1	Knottingley	34.00	30*		2 E	36.30	sigs*	½ L	
32.2	Womersley	39.10	64			-			1/220 F
38.6	Shaftholme Junction	45.25	45*		4½ E	49.00		1 E	L
42.2	Doncaster	50.10	61½		4¾ E	55.55	sigs*	1 L	L
49.3	Piper's Wood	56.55	52½			64.10	48		1/198 R
51.1	Bawtry	58.45	68			66.10	64		1/198 F
54.9	Ranskill	62.10	63			69.40	65		L
60.2	Retford	67.25	58½ /62		6½ E	75.10	57	1¼ L	1/198 R
65.1	Markham Box	72.50	49½			81.10	43		1/178, 200 R
71.4	Crow Park	78.35	76½			87.10	72		1/200 F
78.7	Newark	86.10/86.40	sig stand		6¼ E	93.45	61½	¾ L	L
87.3	Hougham	99.00	54			102.35	56		1/300 R
	Peascliffe Tunnel	-	44½			-			1/200 R
93.3	Grantham	106.35	52½		2½ E	109.55	52	1 L	
98.7	Stoke Box	113.35	41½			116.50	43		1/200 R
101.7	Corby Glen	116.50				120.00			1/178 F
106.6	Little Bytham	120.55	83½			124.05	82		1/200 F
110.2	Essendine	123.30				126.45			1/264 F
114	Tallington	126.25	78/pws			129.40	78		L
119.3	Werrington Junction	132.30				134.10			
122.4	Peterborough	136.10			2¾ E	138.10		¾ E	
129.4	Holme	145.00	61½			147.05	65		L

124 • LONDON & NORTH EASTERN RAILWAY 4-4-0 TENDER LOCOMOTIVES

		Harrogate–King's Cross Pullman					
		5511 *Marne*			**5511** *Marne*		
		6 Pullmans, 248/255 tons			7 Pullmans, 282/290 tons		
Miles	Location	Times	Speed		Times	Speed	Gradients
135.3	Abbots Ripton	151.30	46½		153.10	51	1/200 R
139.9	Huntingdon	156.20	71½	1¾ E	157.55	68	1/200 F
147.1	St Neot's	163.10	54		164.30	58	1/200 R
154.7	Sandy	170.45	65		172.00	64	L
157.7	Biggleswade	173.40	62		174.55	62	
161.8	Arlesey	178.05	56		179.10	58	1/264 R
166.9	Hitchin	183.45	pws 30*	1¼ E	184.40	47	¼ E	1/200 R
170.2	Stevenage	188.45	40		188.55	45	1/200 R
173.8	Knebworth	193.05	52		192.45	57	1/330 F
176.8	Welwyn North	196.25			-		1/200 F
181.1	Hatfield	200.20	71½	¼ L	199.30	70	½ E	1/200 F
186.1	Potters Bar	205.20	57½/ sigs		204.35	53½	1/200 R
189.6	New Barnet	210.30	69		208.00	76	1/200 F
196.2	Finsbury Park	217.00			213.40		
<u>198.8</u>	<u>King's Cross</u>	<u>221.20</u>	<u>(210½ net)</u>	<u>1¼ L</u>	<u>218.00</u>	<u>(215 net)</u>	<u>2 E</u>

5511 *Marne* on test at Gorton, c1925.
(MLS Collection)

5506 *Butler Henderson* passing through Wood Green with the Up 'Harrogate Pullman', c1928. (MLS Collection)

5502 *Zeebrugge* with the *Queen of Scots* Pullman train at Marshmoor, 1930. (MLS Collection)

		Marylebone–Leicester Central, c1928-30								
		4.55pm M'bone 5502 *Zeebrugge* 175/180 tons			4.55pm M'bone 5504 *Jutland* 175/180 tons			3.20pm M'bone 5510 *Princess Mary* 272/285 tons		
Miles	Location	Times	Speed		Times	Speed		Times	Speed		Gradients
0	Marylebone	00.00			00.00			00.00			1/100 R
5.1	Neasden	07.30	70 ½	1 E	08.15	65	¼ E	09.00	66	T	1/90 F
9.2	Harrow-on-the-Hill	11.35	50	1½ E	12.30	pws	½ E	14.40	pws	¾ L	1/91 R
11.4	Pinner	14.35			17.10			17.15			1/176 F
13.7	Northwood	17.45			18.30			19.45	60		1/145 R, 1/176F
17.2	Rickmansworth	21.25	*	1½ E	22.05	*	T	23.15	66/ *	¼ L	
19.4	Chorley Wood	24.35	41 ½		25.40	37		26.30	36		1/106 R
23.6	Amersham	30.25	44		32.55	35		33.35	35		1/105 R
28.8	Great Missenden	35.15	75	¼ L	38.10		3 ¼ L	38.45	69/53	2¾ L	
33.3	Wendover	39.35	76½		42.35	77½		43.15	72½		1/117 F
38	Aylesbury	43.30	60	1½ E	46.30		1½ L	47.20		1¼ L	1/117 F
44	Quainton Road	49.30		2½ E	52.15		¼ L	53.30		½ L	
46.8	Grendon Underwood	52.15	66		54.55	69		56.00			L, 1/176 F
48.8	Calvert	54.05			56.45	64/69		57.50	66		1/176 R
54.5	Finmere	59.45	52		62.10	54		63.30	50		1/176 R
59.3	Brackley	64.55			67.05	68		68.30	66		1/176 F
62.5	Helmdon	68.35	53		70.35	55½		72.00	54		1/176 R
66.1	Culworth	72.15	68		73.55	74		75.25	72½		L, 1/176 F
69.1	Woodford	75.00		1 E	76.30		½ L	78.00		1 L	L
71.5	Charwelton	77.25	65		78.50	64½		80.15			1/176 R
78.5	Braunston	83.15	80½		84.55	77½		86.15	79		1/176 F
83.2	Rugby	87.25			89.10	60		90.20	60		1/176 R
90	Lutterworth	94.00	easy		96.00	69/53		97.05	69/53		1/176 F, 1/176R
93.9	Ashby Magna	97.45			100.00			100.45			1/176 F
98.4	Whetstone	101.40	72		103.55	75		104.15	79		1/176 F
103.1	Leicester	106.30		1½ E	108.30		½ L	108.55		1 E	

The schedule of *The Queen of Scots* allowed 81 minutes to pass Peterborough (76.4 miles) and in July 1930 5506 is recorded as having passed that city in just 74 minutes, averaging 76.5mph over the 27-mile Hitchin–Huntingdon racing stretch with a maximum of 88½mph at Three Counties. The B3s displaced from the East Coast work replaced the D11s at Gorton and some of the Neasden D11s, of which just three, 5501, 5504 and 5505, remained there. By 1933 the allocation of the eleven engines had returned to:

Gorton:	5501 – 5503, 5508, 5509, 5511
Neasden:	5504 – 5507, 5510

In the late 1920s, the D11s continued to run to a continuing high standard on trains like the 3.20pm and 4.55pm Marylebone with their 108/109 minute schedules to Leicester. Here are three more examples from that period.

Professor Tuplin described a fine run he had behind 5510 *Princess Mary* in the spring of 1934 on the 3.20pm from Marylebone. The load was eight coaches including a restaurant car, weighing 276 tons tare and estimated at 295 tons gross as the train was very full. 5510 was blowing off steam furiously before departure and it made more noise blasting out of the station on the 1 in 95/100 to the first tunnel. 67mph past Neasden, falling to 52 on the 1 in 91 to Harrow and 65/59 after Pinner saw the train through Rickmansworth in just over 21 minutes. The engine was worked hard on the climb to Amersham, although the speed fell to 32mph on the 7-mile climb, recovering to 68 in the dip to Great Missenden and falling to 49 on the 1 in 125 to Dutchlands Box. 5510 swept down through Wendover at 72 and 77mph at Stoke Mandeville before braking for Aylesbury passed in 45 minutes from London. Speed varied between 71 and 64 over the next few miles. The 25 miles from Calvert to Helmdon are almost continuously uphill on long stretches of 1 in 176 with just a 2-mile dip after Finmere. The climb started at 70mph, had dropped to 60 by Finmere, recovered to 64 in the dip and cleared the summit at 57mph, a fine climb with this load. Down the long 1 in 176 to Braunston 5510 accelerated to 80mph and later 79 at Whetstone bringing the train to a stand at Leicester in just under 106 minutes, 3 minutes early. The train was held to its advertised starting time of 5.11pm and then ran the 23.4 miles to Nottingham in 24 minutes 34 seconds with 50 at Belgrave, 75 at Loughborough, 57 at Barnston Box, 75 at Gotham Sidings continuing in the 70s to Arkwright Street. 5510 blew off steam whilst standing in Nottingham station and full pressure was needed for the next 11 miles, mostly graded at 1 in 130. Speed hovered around 45mph for most of the climb through the Nottinghamshire coal field and from the summit near Kirkby Bentinck the train was restricted in speed because of colliery area subsidence. Released from the slacks for a moment, 5510 was allowed to reach a top speed of a full 90mph on the falling gradients through Staveley (it averaged 89.6mph from Staveley to Eckington, with the regulator closed, according to the driver!) and Sheffield was reached in 45 minutes 17 seconds against the 48 minute schedule for the 38 miles. Another coach was added to the train at Sheffield Victoria, with the load now a full 360 tons, with 5510 and her driver (and fireman) now facing their stiffest challenge – 13 miles on 1 in 130/120 to Penistone. The minimum speed was 34mph and the train just held the 25-minute schedule. The engine and crew were going through to Manchester, and it was a shame that Tuplin was finishing his run there, as it would be interesting to know how 5510 would cope with the continuing climb to Woodhead Tunnel, but steam pressure had been maintained and despite the fact that the fireman was very tired, there is no reason to think that the schedule could not have been achieved. To make the task a little easier, two coaches had been detached at Penistone for Halifax. This was an arduous double-home turn for the Neasden crew, and they would return with their engine on the morning Manchester–Marylebone the next day. Because of the load, the engine was pushed to the limit on the climbs, but like the other D10s and D11s, ran very freely on the long downgrades easily free-wheeling up to the mid-70s – drifting easily as a result of the fitting of Trofinor valves.

Two runs from Sheffield to Manchester London Road were logged and published in the *Railway Magazine* in 1926 and 1931 and are shown below:

		Sheffield Victoria–Manchester London Road					
		5502 *Zeebrugge* 6 chs, 167/175 tons 4.55pm Marylebone			5501 *Mons* 8 chs, 251/265 tons 3.20pm Marylebone		
Miles	Location	Times	Speed		Times	Speed		Gradients
0	Sheffield Victoria	00.00		T	00.00			1/132 R
1.2	Neepsend	02.45	38		-			1/132 R
2.9	Wadsley Bridge	08.15	45		-	38		1/132 R
4.9	Oughty Bridge	07.50	46		08.55	40½		1/120 R
7.9	Deepcar	11.45	45		13.50	37½		1/120 R
8.8	Wortley	12.55	47½		15.10	39½		1/120 R
12.2	Barnsley Junction	17.35	43½		-			1/131 R
12.9	Penistone	18.40	30*	1¼ E	21.35		1½ E	1/100 R
16.7	Hazlehead Bridge	25.00	36/39		07.45	40	T	1/130 R
18.9	Dunford Bridge	28.05	45	1 E	11.00	43½		1/135 R
22.1	Woodhead	31.45		1¼ E	15.00		1 E	1/201 F
24.1	Crowden	33.40			-			1/117 F
26	Torside Box	35.25	66		18.40	72½		1/117 F
29.4	Dinting	39.05	55*/pws		22.15	40*		1/100 F
33	Godley Junction	44.45	pws		24.45	64		
36.3	Guide Bridge	48.40		1¼ E	30.10		2 E	
40.5	Ardwick	54.25						
41.3	Manchester London Rd	56.20		2¾ E				

Both runs are at the end of a long through working by Neasden engines and men, the latter particularly meritorious with the heavier load. Cecil J. Allen calculated 5501's edhp as 1,015 on the climb to Woodhead, both before and after Penistone. This appears to be the highest recorded horsepower published for a 'Director'.

With the introduction of some B17s onto the GC main line, 5501, 5508, 5509 and 5511 moved to Sheffield, joining the allocation of GN Atlantics there, including the nightly turn via Woodford and Banbury to Swindon. This working appears to have been shared with GWR engines on a 2-year rota basis. At the end of the LNER coverage in 1935, these Sheffield engines returned to Gorton. The D11s continued to work to London despite the availability of the B17s which were initially not too popular and it was only in 1936/7 that they were displaced by the B17/4s (the 'Footballer' 4-6-0s) which were specifically ordered for the Great Central lines.

5510 *Princess Mary* with a Marylebone–Manchester express at Dunford Bridge, c1929. (H. Gordon Tidey/MLS Collection)

As a result, the D11s (and also the D10s) were regrouped at Sheffield to take over the extensive work there previously allocated to the D9s. 5501 *Mons* was timed between Liverpool and Manchester on the 4pm to Hull with seven coaches, 215 tons gross, and kept the 45 minutes schedule including station stops at Farnworth and Warrington and a p-way restriction between Urmston and Trafford Park. It topped the 1 in 195 to Hunt's Cross at 48mph and 69mph in the dip afterwards before the Farnworth stop. 63½mph was attained down the 1 in 158 to Sankey on the 8¾ minute run to Warrington and 65mph before and 68mph after the rise over the Manchester Ship Canal, which was cleared at 57½mph. Then, just a year or so before the Second World War, 5503 *Somme* was in

5511 *Marne* passing Guide Bridge with the 3.20pm Marylebone–Manchester, 6 August 1932. (MLS Collection)

5505 *Ypres* departing from Guide Bridge with a Manchester–Marylebone express, 19 September 1933. (R.D. Pollard/MLS Collection)

5501 *Mons* arriving at Manchester London Road with the 8.45am Marylebone just before the 'Directors' were displaced from the main services by the Gresley B17/4 'Football' 4-6-0s, 30 April 1936. (MLS Collection)

charge of another similarly loaded Manchester–Liverpool train and most unusually had almost achieved even time passing Padgate Junction, 14.2 miles in 14 minutes 52 seconds with speeds of 70½mph at Flixton, 60 over the Manchester Ship Canal and 72½mph just before Padgate Junction and a signal stand outside Warrington, reached in consequence a minute late. This was more than regained with 56 minimum of the 1 in 158 after Sankey, arriving at Farnworth half a minute early. Despite a dash through Mersey Road at 67mph, arrival in Liverpool was three minutes late caused by severe signal checks at Garston and St Michael's. This was a lot more sprightly than when some of the D11s were stabled there in their declining days.

At the outbreak of war in 1939, only 5503 and 5511 remained at Gorton. The Neasden D11s, apart from 5506, remained there throughout the war (5506 had gone to Sheffield in 1938) working the 4am newspaper train and semi-fast services to Woodford and Leicester. Sheffield's 5508 went to Lincoln in 1939 and on to Immingham in 1940. Langwith received 5503 and 5511 in 1942 and 5501 in 1943, while Mexborough gained 5502 and 5506 in 1943 and the Langwith engines in later in 1943 and 1944. The heaviest war work was on expresses between Sheffield and Hull. Other work included work at the Royal Ordnance Factory at Ranskill, and stopping trains in the Sheffield, Doncaster, Lincoln area. At the end of the war in 1945 the distribution was:

Immingham:	5508, 5509, 5511
Neasden:	5504, 5505, 5507, 5510
Mexborough:	5501–5503, 5506

5508 *Prince of Wales*, displaced from top-link work, at Vale House with the 11.22am Manchester Central–Barnetby, 2 August 1937. (MLS Collection)

5502 *Zeebrugge* near Dinting on the 3.52pm Manchester Central–Sheffield Victoria stopping train, 2 August 1937. (R.D. Pollard/MLS Collection)

By 1947, the whole class was based at Immingham, the most prestigious turn being the Harwich–Liverpool boat train between March and Sheffield, a turn they retained until 1950. In the spring of 1950, 62663, 62665 and 62670 were transferred to the CLC at Trafford Park and Walton, then 62663 and 62665 at Stockport, the latter two working both express and stopping trains between Manchester and Liverpool. Later in 1950 62660–62662 and 62666–62670 were transferred to Trafford Park, though most were in store. Before this transfer took place, I, then a 12 year old schoolboy spending the summer with a school friend in Doncaster, remember seeing 62661 *Gerard Powys Dewhurst* and 62668 *Jutland* arriving at Doncaster with stopping trains from Sheffield. In April 1951 only the two Stockport engines and 62664 were in use. 62669 joined the active D10s at Northwich in 1952.

In 1953, all bar 62660 and 62668 came out of store and the two remaining Immingham engines, 62666 and 62667, went to Mexborough for the summer service, moving to Lincoln in the autumn. 62660 later joined them at Lincoln. The Lincoln engines were used on stopping trains to Nottingham and Derby, retaining this work until replaced by diesel multiple units which led to their withdrawal. The two Stockport engines were transferred in 1954, 62663 to Lincoln and 62665 to Northwich. The allocation in 1957 was:

Lincoln:	62660, 62663, 62666, 62667, 62670
Northwich:	62661, 62662, 62664, 62665, 62669
Trafford Park:	62668 (in store)

The Lincoln engines were transferred to Sheffield Darnall in March 1957 and shared work with B1s to Doncaster and York as well as local stopping services though in the 1957/8 winter service all of these were stored. The Northwich and Trafford Park engines were

62670 *Marne* passing Flixton on the 2.42pm Harwich–Liverpool boat train, 19 April 1952. (J.D. Darby/MLS Collection)

moved to Darnall in April 1958, all apart from 62663 at Staveley and 62662 (which acted as station pilot at Sheffield Victoria) in store. The first to be withdrawn in May 1959 was 62665, but the rest were taken out of store for the 1959 summer service and were used mainly on Sheffield–Nottingham slow trains, releasing B1s for the holiday traffic to the East Coast resorts. 62661 joined 62663 at Staveley, the eight Darnall D11s going into store again at the end of the summer service, until the following year when again they emerged from store. Four were withdrawn in August 1960, the remainder going by the end of the year. The last was 62666, though 62660 was saved for preservation (see page 141).

During their final years, several were used for enthusiast railtours. 62667 *Somme* was used in June 1953 for two RCTS tours. Alan Pegler of the Northern Rubber Company charted a special from Retford to Bourne End for a Thames River cruise staff outing, using 62666 *Zeebrugge*. It had a substantial 11-coach 400 ton gross load and was worked by a Retford crew throughout with suitable pilotmen. After threading the Nottinghamshire coalfields, it left Nottingham on time, fell from 43 to 36mph on the 1 in 176 to East Leake and just touched 60mph at Loughborough. The last calling point was Leicester, left 6 minutes late. 62666 accelerated to 48½ by Whetstone and gradually fell back to 36mph on the long 1 in 176 beyond Ashby Magna. 65mph was reached at Braunston and Rugby was passed 5 minutes late. The long drag up six miles of 1 in 176 through Catesby Tunnel to Charwelton brought the speed down to 38½mph and a top speed of 67mph was reached at Braunston. Lateness was under 2 minutes when it left the GC main line at Grendon Junction and had been eliminated on reaching the GW/GC joint line at Ashendon Junction. Saunderton summit was cleared at 42mph and a final 60 brought the excursion into High Wycombe 3 minutes early after its 3½ hour run. In September, Pegler ran a special to the Farnborough air show, double-headed by the preserved GNR Atlantic 251 and D11 62663 *Prince Albert* and a trip to Blackpool. In 1956 the Ian Allan organisation ran a special 'Pennine Pullman' hauled by 62662 *Prince of Wales* and 62664 *Princess Mary* from Manchester Ardwick, where they replace Co-Co electric 27002, to Barnsley, Mexborough and Rotherham, where 60014 took over for the return to King's Cross.

62663 *Prince Albert* is joined by preserved GN C1 Atlantic 251 for a special Northern Rubber Plant Works train organised by Alan Pegler, seen passing Manchester Exchange, 4 September 1954. What seems to be an unrebuilt 'Patriot' is passing in the opposite direction.
(H.D. Bowtell/MLS Collection)

62667 *Somme* pulling out of Lincoln St Marks with the RCTS 'South Yorkshire' railtour, 7 June 1953. (P. Ward/MLS Collection)

62664 *Princess Mary* and 62662 *Prince of Wales* with the *Trains Illustrated* 'Pennine Pullman' at Luddenfoot, 12 May 1956. (B.K.B. Green/MLS Collection)

62662 *Prince of Wales* on a Cheshire Lines Committee local train near Ashley, 15 October 1955. (G.M. Shoults/ MLS Collection)

62664 *Princess Mary* at Chester Northgate station, 6 June 1956. (MLS Collection)

62668 *Jutland* on a summer special Sheffield – Cleethorpes train leaving Retford, c1956. (J. Davenport/ MLS Collection)

62661 *Gerard Powys Dewhurst* at Retford with a Sheffield–Lincoln stopping train, 2 August 1958. (A.C. Gilbert/ MLS Collection)

62662 *Prince of Wales* at Waleswood with the 9.30am Manchester–Cleethorpes summer holiday express, 2 August 1958. (A.C. Gilbert/MLS Collection)

62665 *Mons* with a CLC local train near Flixton, 1958. (G.M. Shoults/MLS Collection)

The Great Central 4-4-0s • 139

A busy scene at Sheffield Victoria with 62662 *Prince of Wales,* Ivatt 4MT 2-6-0 43058, a V2 and an O4 approaching with mineral empties, summer 1958. (J.H. Turner/MLS Collection)

62670 *Marne* with a Down parcels train at Staveley, 22 August 1959. (MLS Collection)

In a 1956 *Railway Magazine* article, Cecil J. Allen compared the power outputs of various British locomotives and placed the top 40 in ascending order based on their drawbar horsepower produced per foot of grate area. For comparison he noted that none surpassed the French Chapelon 4-8-0s and Pacifics, and a Kylchap A4, 4901, with its performance on a 730 ton East Coast wartime express, topped the British locomotives. Surprisingly a GW mogul (5326) came next with a phenomenal effort on a 465 ton train out of Leamington followed by a GW 'Star' and 'King'. The GW 'Saints' and 'Castles' also featured in the top ten (the Kings and Castles before fitting with high superheat and double chimneys). The highest placed 4-4-0 was a LNWR 'George V' at '5'. A GNR Atlantic was placed seventh. The top North Eastern engine was a 'Z' Atlantic at '13' and the LNER 'B17' came in twentieth. The GE 'Claud Hamilton' 'D16' was No. 22 and the highest GC engine was the D11 'Director' at twenty-six. The only other LNER 4-4-0 that figured in the top 40 was the 'D49' at thirty-five. The LNER 'A3' Pacific was '24' and the 'P2' 2-8-2 twenty-fifth. These assessments were based on logs sent to the *Railway Magazine* over the previous fifty years. He did not assess the coal consumption required to achieve these high performance figures, but there is no doubt from other published figures that the GW engines were by some way the most economical. The GCR did not publish openly any such figures, although their 4-6-0s were said to be requiring 80lbs per mile, a very excessive use of fuel. The GW 'Stars' and 'Castles' were half that and it is thought that the 'Directors' were in-between at around 60lbs per mile on main line work.

Preservation – 506 *Butler-Henderson*

506 *Butler-Henderson* was constructed in December 1919 at a

62660 *Butler-Henderson* at Sheffield Victoria with an RTCS railtour, 21 September 1958. (Real Photos/MLS Collection)

cost of £7,620 and was named after the director appointed to the GCR Board in 1918. It was initially based at Neasden, but in June 1925 moved to Gorton and then in April 1927 at Copley Hill Leeds from whence it shared the work on the Pullman trains to King's Cross. It moved back to Neasden in 1928, where it remained for the next ten years. Displaced by the B17s in 1938, it moved to Sheffield Darnall, then Mexborough and at nationalisation to Immingham. In February 1951 it was transferred to the CLC lines at Trafford Park, in November 1953 to Lincoln and its final depot was Darnall in November 1957. It was withdrawn in October 1960, having run a total of 1,280,897 miles in traffic in its 38 years of operation.

It was then selected for preservation and stored at Gorton awaiting restoration, which was carried out at Gorton Works in 1963. It was finished in its original Great Central livery, the valances over the coupled wheels restored, and placed in the BTC Museum in the former bus depot at Clapham. In 1975 it was placed in the national collection and loaned to the Great Central Heritage Railway, where it was restored to operational activity in 1981/2 and was in steam on the GCR from March 1983. At the end of its boiler certificate, it was briefly repainted in BR mixed traffic lined black as 62660, before withdrawal, repainting in GCR livery and renumbering 506 and placed as a static exhibit at the National Railway Museum at York. In 2005 it joined the Midland Compound and other preserved engines at the Roundhouse at Barrow Hill where it currently remains in 2023.

62660 at Gorton awaiting restoration, 18 March 1961. (MLS Collection)

The restored 506 *Butler-Henderson* at Gorton Top Yard after restoration, 22 July 1963. In front of 506 are four retired GC drivers, from left to right, Vinson Gulliver, Bert Wagstaffe, Unknown and Arthur Davies. Vinson was still alive in 1996 aged 107, the oldest railwayman in Europe! (R.E. Gee/MLS Collection)

506 being hauled to Romily for official photos at Priory Junction, 24 July 1963.
(R.E. Gee/MLS Collection)

506 in the BTC transport Museum at a former bus depot at Clapham alongside Midland Compound 1000, c1970.
(MLS Collection)

The D11/2, J.G. Robinson/Nigel Gresley, 1924

The last two Great Central '11Fs', 505 and 511, reclassified by the new LNER Company as class D11, were only constructed in December 1922, days before they were taken into the stock of the new company. In fact, 511 was actually painted in the new LNER livery of lined green without the Great Central Railway coat of arms on the tender. Nigel Gresley, the Great Northern Locomotive Superintendent, took over the new LNER Chief Mechanical Engineer's role, though he retained Robinson, the former GCR man, as a consultant on locomotive matters. In the first couple of years Gresley and Doncaster Works were preoccupied with the design and construction of the new A1 Pacifics, but there was some concern with the locomotive availability in Scotland, especially for the key services radiating from Edinburgh to Glasgow, Dundee and Aberdeen. The North British had constructed no recent main line locomotives and so Gresley, no doubt with advice from Robinson, placed orders for twenty-four of the most recent successful medium-sized design with Kitson and Armstrong Whitworth – the GC D11s.

Being required quickly, both companies were under pressure and Kitson delivered 6378–6389 between July and October 1924, and Armstrong Whitworth constructed their dozen, 6390–6401 in October and November of that year. Minimum alterations were made to the design, the most significant being the need to reduce the height of the locomotives to fit the North British loading gauge rather than the more generous GCR one. They therefore had lower cabs and boiler mountings and their tenders were not equipped with scoops as the former NBR possessed no water troughs. The new engines were finished in the LNER lined apple green livery, were classified as sub-group D11/2, and were named after characters from Sir Walter Scott's novels, a theme used previously on some of the NB 4-4-0s. The names for the Scottish engines were:

6378	*Bailie MacWheeble*
6379	*Baron of Bradwardine*
6380	*Evan Dhu*
6381	*Flora MacIvor*
6382	*Colonel Gardiner*
6383	*Jonathan Oldbuck*
6384	*Edie Ochiltree*
6385	*Luckie Mucklebackit*
6386	*Lord Glenallan*
6387	*Lucy Ashton*
6388	*Captain Craigengelt*
6389	*Haystoun of Bucklaw*
6390	*Hobbie Elliott*
6391	*Wizard of the Moor*
6392	*Malcolm Graeme*
6393	*The Fiery Cross*
6394	*Lord James of Douglas*
6395	*Ellen Douglas*
6396	*Maid of Lorn*
6397	*The Lady of the Lake*
6398	*Laird of Balmawhapple*
6399	*Allan-Bane*
6400	*Roderick Dhu*
6401	*James Fitzjames*

To meet the loading gauge the locomotives were fitted with the Gresley one foot high 'flowerpot' chimney with the flattened dome, shorter pattern of Ross pop safety valves and lower cab. The brass beading around the splashers was omitted. 6386 and 6393-6396 were fitted with tablet catching apparatus on their tenders for working from Dundee shed, although this was removed when they were transferred elsewhere. The valances under the raised portion of the running plate were removed from 1925 onwards. Drop grates were fitted to eight engines – 6378-6380, 6387-6389, 6397 and 6398 – between 1933 and 1935 and the remainder were so equipped after the war. Some of the engines exchanged their smokebox doors with those from the NBR Atlantics and some acquired doors with curved handrails. Some were equipped with Strowger-Hudd ATC gear after the Castlecary collision in 1937, but although the start was made in 1939, with eleven fitted, the war priorities interrupted the application and the gear was removed from these in 1942. Like

6379 *Baron of Bradwardine* at Glasgow Eastfield, 27 September 1925. It is as delivered by Kitson and painted in LNER lined green livery, reduced boiler mountings including 'flowerpot' chimney and still with the valance above the coupled wheels. (W. Potter/MLS Collection)

the former GC 'Directors' the Scottish engines reverted from lined green to black in 1928 but after the war sixteen of these were restored to lined apple green and passed to BR still in that livery – the engines, now renumbered in the 1946 scheme, 2671, 2677, 2679-2681, 2683-2688, 2690-2694. Four ran with BR numbers in LNER green but 'BRITISH RAILWAYS' on the tender – 62671, 62677, 62683 and 62690. 62671 lasted in this livery until 1953. Some were turned out in BR Mixed traffic lined livery, others in plain black after 1948. The first withdrawals were in September 1958 (62679 and 62683). The last three were 62691 and 62693 in November 1961 and 62685 in January 1962.

6385 Luckie Mucklebackit as built and delivered by Kitson & Co., seen here c1926. (MLS Collection)

6394 Lord James of Douglas at Eastfield, May 1937. It was built by Armstrong Whitworth in November 1924 and is seen here after the valance above the coupled wheels has been removed and the lined green livery replaced by plain black. (W. Potter/MLS Collection)

6388 *Captain Craigengelt* coupled with a D49/1 'Shire' about to go off Glasgow Eastfield shed, 1932. (J.A.G. Coltas/MLS Collection)

6382 *Colonel Gardiner* in typical post-war rundown condition, c1945. (Photomatic/MLS Collection)

2694 *James Fitzjames* also in LNER lined green and with a smokebox door off a former NBR Atlantic with curved handrail, at Thornton, 7 October 1948.
(J.D. Darby/MLS Collection)

62692 *Allan-Bane* at Dundee in BR mixed traffic lined black livery, 1 September 1952.
(N.R. Knight/MLS Collection)

62686 *The Fiery Cross* and 62673 *Evan Dhu* at Glasgow Eastfield shed, both in BR lined black livery, 7 June 1953. (A.C. Gilbert/MLS Collection)

62678 *Luckie Mucklebackit,* one of the Scottish Director to be withdrawn early, with a J36 0-6-0 and a D49/1 at Thornton Junction, 30 July 1959. It had been withdrawn in March of that year. (MLS Collection)

A number of Directors were stored for long periods at both Polmont and Longniddry in the late 1950s. Here are 62693 *Roderick Dhu* and 62689 *Maid of Lorn* at Polmont, 25 July 1960. 62689 was withdrawn in July and 62693 in November 1961.
(MLS Collection)

Operation

The initial allocation at the end of 1924 was:

Glasgow Eastfield:	6378–6380, 6397, 6398, 6400
Edinburgh Haymarket:	6381–6385, 6401
Edinburgh St Margaret's:	6389, 6392, 6399
Dundee:	6386, 6393–6396
Perth:	6387, 6388

There were no changes to this allocation until the building of the Gresley D49 'Shires' and the allocation of the first batch to

Glasgow Queen Street–Edinburgh Haymarket, 1924

		6379 (Eastfield) *Baron of Bradwardine* 326/345 tons 8.40am Glasgow		6379 (Eastfield) *Baron of Bradwardine* 330/350 tons		6382 (Haymarket) *Colonel Gardiner* 330/360 tons		
Miles	Location	Times	Speed	Times	Speed	Times	Speed	Gradient
0	Glasgow Queen St	00.00		00.00		00.00		1/41 R
1.5	Cowlairs	07.00	*banked	06.05	*banked	06.39	*banked	1/41 R
6.2	Lenzie	14.25	53 1½ L	13.40/14.10	sigs	13.54	pws T	L
11.4	Croy	20.10	65	22.35		20.15		1/900 R
15.5	Castlecary	24.10	61	26.30	65	24.45	59	L
21.8	Falkirk	30.35	59/62/pws*	32.25	69/pws	31.00	65/32*	L
<u>25</u>	<u>Polmont</u>	<u>36.06</u>	2 L	<u>37.05</u>	3 L	36.16	pws	
0		00.00		00.00				
4.7	Linlithgow	07.20		07.20		41.46	54½	1/882 R
7.8	Philipstoun	10.45	55	10.40	56½	-	51½	1/882 R
10.3	Winchburgh Junction	13.25	58	13.15	61½	48.14		1/960 F
14.1	Ratho	17.25	64	17.10	67	52.03	65	1/960 F
18.8	Saughton Junction	21.45	66	21.05	72/ sigs	56.33	62	1/960 F
<u>21.1</u>	<u>Haymarket</u>	<u>24.25</u>	½ L	<u>24.30</u>	1½ L	<u>59.19</u>	¾ E	

Scotland in the late 1920s. The *Railway Magazine* took an interest in their arrival in Scotland and in January 1925 published the description of a series of runs between Glasgow and Edinburgh Haymarket. They were all with substantial loads on the 60 minute schedules, one non-stop and with stops at Polmont only.

Another logged run featured 6398 *Laird of Balmawhapple* with 240/265 tons which was badly delayed by two p-way restrictions and two signal checks before Polmont. It left there 3½ minutes late and with 70mph at Ratho and 67 at Saughton Junction, regained 2½ minutes to Haymarket until checked between Saughton Junction and Haymarket. As a result, it terminated still 2½ minutes late. Later, in January 1936, 6400 *Roderick Dhu* with nine coaches, 300/315 tons, got to Polmont in 30 minutes 48 seconds gaining just ¼ minute on the accelerated 31 minute timing with speeds of 56mph at Lenzie, 66 at Greenhill Upper Junction and 69 before Falkirk. After a p-way slowing to 40mph at Winchburgh Junction, 6400 accelerated hard to achieve 64mph at Ratho and 70 at Saughton Junction, passing Haymarket in 24¾ minutes from Polmont and 27½ to a stand at Edinburgh Waverley just ½ minute late.

In the opposite direction 6401 *James Fitzjames* had charge of the *Flying Scotsman* train on its final leg from Edinburgh to Glasgow Queen street. The date was in 1927.

The Perth engines were transferred to Eastfield in 1928 and a couple of Eastfield engines went to Haymarket. Initially the Eastfield and Haymarket engines were allocated to regular crews and worked the main trains between Glasgow and Edinburgh, including the *Queen of Scots* Pullman train, and trains to Arbroath, Dundee and Aberdeen. They worked rarely south of Edinburgh although they did work to Newcastle on a meat train that ran irregularly or as pilot to a Pacific. There was a brief experiment sending 6399 to Stratford in November 1926 to see if it was an answer to motive power problems on the former GE section, but because of its axleload, its route availability would have been restricted and extra K2 moguls were provided until more B12s had been constructed. Before returning to Scotland 6399 was tested on Leeds–King's Cross express services prior to the allocation of former GC Directors for the Harrogate/Leeds–London Pullman services. I now table a couple of runs from Dundee to Edinburgh Waverley with Aberdeen–Edinburgh expresses, the first shortly after 6384's construction in 1924 and the second with a pair of D11s on a 500 ton train in 1934 – pairs of 4-4-0s were used unless a P2 2-8-2 or A3 was available on the heaviest Edinburgh–Aberdeen main services.

Edinburgh Waverley–Glasgow Queen Street

6401 *James Fitzjames* (Haymarket)

355 tons

10am king's Cross, *The Flying Scotsman*,

Miles	Location	Times	Speed	Gradients
0	Edinburgh Waverley	00.00		
1.2	Haymarket	03.13		
5.5	Gogar	09.06		1/960 R
12	Winchburgh Jcn	16.20	54	1/960 R
17.6	Linlithgow	22.20	56	1/882 F
22.3	Polmont	27.17		
0		00.00		
2.2	Falkirk	06.12		1.682 R
6.6	Bonnybridge	10.06		L
9.5	Castlecary	13.13	56	L
12.1	Dullatur	16.00		L
13.6	Croy	17.28	66/ pws	1/900 F
18.7	Lenzie	23.21		L
23.5	Cowlairs	29.13		
25.1	Glasgow Queen St	34.55	(32½ net)	

		Dundee–Edinburgh Waverley (trains ex Aberdeen)					
		6384 (Haymarket)			6385 (Haymarket) + 6395 (Dundee)		
		Edie Ochiltree			*Luckie Mucklebackit & Ellen Douglas*		
		278/295 tons			500 tons		
		1924			1934		
Miles	Location	Times	Speed		Times	Speed	Gradients
0	Dundee Tay Bridge	00.00			00.00		1/74, 114 R
2.7	Tay Bridge (South)	07.45		¼ E	07.34		½ E	1/130 R
4.6	St Fort	10.10	45/65		10.14	64		1/100 F
8.3	Leuchars Junction	14.10	67	¼ L	14.00	69	T	L, 1/100 F
11.4	Dairsie	-	52		-	54½		1/160 R
14.6	Cupar	20.55	64		20.36	60		1/400 F
16.9	Springfield	23.25	51½		23.10	54½		1/161 R
20.1	Ladybank Junction	27.05	58	1 E	26.40	60	1 ¼ E	
24.3	Lochmuir Box	33.05	32		32.20	34½		1/95, 105 R
28.5	Thornton Junction	38.55	10*	T	38.05	10*	1 E	1/132 F
30.2	MP 29 (Dysart)	41.50	34½		-	33		1/131, 158 R
33.3	Kirkcaldy	45.25	70		44.48	71½		1/100, 145 F
36.4	Kinghorn	51.25	pws 5*		47.43	pws 30*		L
39	Burntisland	54.30		1½ L	51.37	35*	1 ¼ E	1/128 F
43.2	Dalgetty Box	61.00	36		57.43	40		1/100 R
46	Inverkeithing	64.55	30*	2 L	61.07	30*		1/94 F
48	Forth Bridge North	69.20	20	2¼ L	65.10	24½	1¾ E	1/70 R
49.7	Dalmeny	72.40		2¾ L	68.25		1½ E	L
52.7	Turnhouse	76.05			71.40			1/100 F
55.7	Saughton Junction	79.05	60		74.58	55		L
58	Haymarket	81.40	sigs	1¾ L	77.12	sigs	2¾ E	
<u>59.2</u>	<u>Edinburgh Waverley</u>	<u>85.05</u>	(80 net)	<u>2 L</u>	<u>81.48</u>	(79 net)	<u>1¼ E</u>	

The St Margaret's D11s worked to Perth and Dundee and occasionally over the Waverley route to Carlisle. The Perth engines worked to Edinburgh and the Dundee engines in both directions to Aberdeen and Edinburgh. This included assisting other classes on heavy Edinburgh–Aberdeen trains and the fast fish train work. The arrival of the D49s and more Pacifics in Scotland displaced the D11s from some of their main express work, but they still shared express work on the Edinburgh–Glasgow and Edinburgh–Dundee lines until the arrival of the Thompson B1 4-6-0s after the Second World War. 6401 had a brief exchange in 1943 with the Thompson rebuilt 395, the 2-cylinder D49, (Leeds Neville Hill/

6394 *Lord James of Douglas* and a North British D25 leaving Aberdeen past Ferryhill with the 2.30pm to St Pancras, 16 September 1927. (K. Nunn/LCGB/MLS Collection)

St Margaret's 6389 *Haystoun of Bucklaw* climbing the 1 in 70 gradient from Inverkeithing to the Forth Bridge with a Dundee–Edinburgh stopping train, c1927. (MLS Collection)

6386 *Lord Glenallan* passing Prince's Street Gardens with the 9am Arbroath–Edinburgh Waverley, 22 June 1936. (MLS Collection)

Haymarket) to enable the rebuilt engine to be tested for its suitability in Scotland. The experiment lasted just three months and 6401 returned to Scotland and 365 to Leeds.

During the war, the D11s, concentrated at Eastfield, St Margaret's and Haymarket, like most other classes, became common user and worked almost indiscriminately in any direction from Edinburgh.

The displaced Eastfield engines worked regularly to Dundee and the Fife Coast, as did the Haymarket engines. In the 1950s some were allocated to Thornton Junction and Dunfermline working in the Edinburgh–Stirling–Thornton area on slow and semi-fast services. I spent a fortnight on holiday in Scotland in July 1958 and on a visit to Edinburgh, managed to get a brief run behind 62677 *Edie Ochiltree* over the Forth Bridge to Inverkeithing on a Dundee stopping train. It was said to be at the time only one of the three not in store, the others being 62678 and 62679, the latter being withdrawn that month. I also visited Longniddry and photographed (in pouring rain) a line of D11s and C16 4-4-2Ts in store. During the late 1950s, most of the class spent long periods in store only being used during the brief summer season. Their official allocation almost throughout the 1950s (although when in store they could be parked elsewhere) was:

Glasgow
  Eastfield: 62671–62676, 62680–62682, 62684, 62686–62689
Edinburgh
  Haymarket: 62677–62679, 62683, 62685, 62690–62694

62677–62679 were reallocated to Thornton Junction in May 1957 and remained there until withdrawal. 62689 and 62693 were allocated briefly to St Margaret's in February 1960 but had returned to Eastfield and Haymarket respectively by July. At the end of the 1960 summer timetable the following were noted in store:

Grangemouth: 62671
Parkhead: 62672, 62680, 62681
Thornton: 62674, 62686, 62687
Polmont: 62682, 62689–62691, 62693
Haymarket: 62685

Withdrawals took place in 1958 and 1959, then a pause until 1961, but as noted above, most spent the intervening time in store. The reason they were not withdrawn earlier is unknown, as many went straight from store to withdrawal. Perhaps the unreliability of some of the early diesels sent to Scotland, especially the North British built D 61XX, indicated to the Scottish motive power management the need to keep some steam locomotives in reserve.

62678 *Luckie Mucklebackit*, recently ex-works in the BR lined black mixed traffic livery, leaving Edinburgh Waverley with a stopping train for Dundee, 15 September 1950.
(G.M. Shoults/MLS Collection)

62674 *Flora MacIvor* at Bishopbriggs with a Edinburgh–Glasgow stopping train, 15 April 1950.
(A.G. Ellis/MLS Collection)

62691 *Laird of Balmawhapple* at Perth with a local train for Dundee, 1 June 1953.
(MLS Collection)

A 'Director' on the West Highland line – 62682 *Haystoun of Bucklaw* with a freight near Arisaig, c1955. (MLS Collection)

A pair of D11s, 62679 *Lord Glenallan* and 62682 *Haystoun of Bucklaw* await their turn at Edinburgh Waverley, 7 July 1956. The leading engine is sporting express headlamps and both have been polished up suggesting they are booked for a special working. A reversed headboard is in front of the smokebox of 62679. (MLS Collection)

One of the last Scottish D11s still operating in 1958, 62677 *Edie Ochitree*, that the author caught on an Edinburgh–Dundee stopping train at Inverkeithing, 12 July 1958. (David Maidment)

A Bachmann model (catalogue 31-135) of D11/2 62690 *Lady of the Lake*, 2012. (David Maidment)

The D12, Charles Sacré, 1877

We now move back in time to the 4-4-0 creation of Charles Sacré, Locomotive Superintendent of the Manchester, Sheffield and Linclnshire Railway from 1859 to 1886. In the 1870s the first 4-4-0s appeared on a number of railways as speeds increased and the riding and stability of 2-4-0s became of concern to some engineers. Initially, longer frames with bogies were required without taking advantage of the extra length to increase the boiler size and the twenty-seven built by Sacré between 1877 and 1880 had a long platform covering the bogies with boiler well set back. The locomotives were double-framed and their key dimensions at the Grouping were:

Cylinders (2 inside):	17½ x 26in
Coupled wheel diameter:	6ft 3½in
Bogie wheel diameter:	3ft 3½in
Stephenson motion with slide valves	
Boiler pressure:	140lbs psi
Heating surface:	1,016sq ft (1,113sq ft after reboilering)
Grate area:	15.58sq ft (15.64sq ft after reboilering)
Axleload:	17 tons 12 cwt
Weight (Engine):	43 tons 19 cwt
(Tender):	25 tons 14 cwt
(Total):	69 tons 13 cwt
Water capacity:	1,800 gallons (the 1880 build with 2,500 gallons)
Coal capacity:	2¾ tons
Tractive effort:	12,550lbs

The new locomotives constructed at Gorton in 1877 were numbered 423–434 and a further ten in 1878, 435–444. Two more, 445 and 446, were built in 1879 and a final three, 4, 128 and 128, in 1880, with the 2,500 gallon tender. They had already lost their smokebox wing plates well before reboilering in the 1890s. The original Smith simple vacuum brake system was found to be inadequate and a decision to replace by the standard automatic vacuum brake in 1881 was endorsed when a brake failure after a parting of the train caused a catastrophic accident in 1884, although it was 1887 before all had been modified.

Sacré 4-4-0 438 at New Holland in March 1886, in the livery of the MS&LR. (Real Photographs/ MLS Collection)

Sacré 4-4-0 430 in Great Central livery, c1900. (F. Moore/MLS Collection)

The Leicester allocated Sacré 4-4-0 442B at Leicester GC shed, 1922. (Locomotive & General/ MLS Collection)

Between 1912 and 1914 the locomotives were renumbered on the GCR duplicate list, adding a 'B' after the number and a few had tenders fitted with weatherboards giving the crew some protection when running tender first.

They survived the First World War, but 424B, 426B and 427B were withdrawn in 1919, six more in 1920, two in 1921 and four in 1922. Twelve became LNER stock as class D12 in 1923, but five were withdrawn that year, the remainder being allocated LNER numbers 6460, and 6463–6468. Three were withdrawn in 1925, three in 1926 and the last one, 6464 (formerly 442B) not until 1930.

Sacré 4-4-0 6464 (formerly 442B) at Bulwell on the Annesley Dido staff train, c1925. (MLS Collection)

6466 (ex 430B) at Annesley, 30 August 1926. It was renumbered in May 1925 and withdrawn in October 1926. Note the weatherboard on the tender. (K.Nunn/LCGB/MLS Collection)

Operation

They initially worked the Manchester, Sheffield and Lincolnshire Railway trains from Manchester to King's Cross as far as Retford where they handed over to GN locomotives, and also the Liverpool–Hull services. From 1883, the engine change point of the London trains was altered to Grantham. However, around this time they were replaced on those trains by Sacré 2-2-2s with 7ft 6in driving wheels introduced in 1882. After the introduction of Parker 4-4-0s (LNER D7 and D8) they worked almost entirely on the CLC Liverpool–Manchester services and local trains in Lincolnshire. The CLC Manchester–Warrington section on the fast services were allowed only 18 minutes for the 15.7 miles, an average of 52.5mph, a scheduled time that remained into late LNER timings with much more powerful locomotives. In the 1890s the loads would, however, seldom have rarely exceeded 120-150 tons. I can trace no records of performance by these locomotives, as they had long been replaced on express workings before train logs were published in railway literature.

Five locomotives – 425B, 428 B, 429 B, 442 B and 444B – were loaned to the Great North of Scotland Railway in 1920 to augment their motive power which had been overworked by the heavy military traffic during the war and was then requiring heavy repair with too many unavailable for traffic. They were returned in January 1921, having undertaken only restricted working as they were not fitted with Westinghouse brakes.

The allocation of the locomotives before the Grouping in 1921 was:

Gorton:	5
Trafford Park:	2
Stockport:	2
Annesley:	2
New Holland:	2
Immingham:	2
Liverpool Brunswick	1
Walton:	1
Leicester:	1

The Gorton engines were used mainly on goods traffic, the Annesley engines worked the local Dido staff train and the Leicester engine (442 B) was apparently only fit for shunting. It was overhauled in 1922, transferred to Annesley for the Dido working and survived as the last member of this class on these duties until March 1930, as 6464.

431 on the CLC between Liverpool and Manchester with a lightweight stopping train, c1912. (F. Moore/MLS Collection)

424B at Manchester Central with a CLC train, c1912. (Real Photos/MLS Collection)

443B with a freight train from Liverpool Brunswick Dewsnap Goods Depot approaching Trafford Park, c1922. It was allocated the number 6463 but this was not worn before withdrawal in February 1925. (MLS Collection)

Chapter 4
THE GREAT EASTERN 4-4-0s

James Holden's
T19 2-4-0 784, later renumbered 773 and withdrawn in 1909 before possible rebuilding as a 4-4-0.
(F. Moore/MLS Collection)

The GE 'T19 RBT' (LNER D13)
James Holden designed and had constructed 110 T19 2-4-0 express engines between 1886 and 1897, numbered 700–779 and 1010–1039. A number of them were oil-burning including 760 which was named *Petrolea*.

761 was the Great Eastern's 'royal engine' for many years, including taking the future King George V and his wife on their honeymoon to Sandringham in 1893. Twenty-one of them were rebuilt with larger high pitched Belpaire boilers between 1902 and 1904 and with their somewhat ungainly top-heavy appearance, were nicknamed 'Humpty Dumpties'. Twenty-nine of the original 2-4-0s were scrapped between 1908 and 1913 and the reboilered engines between 1913 and 1920.

The Great Eastern 4-4-0s • 163

T19 oil burning 2-4-0 760 *Petrolea* as built in 1890, rebuilt later with a higher pitched Belpaire boiler, but withdrawn in 1914 before being rebuilt as a 4-4-0. (MLS Collection)

Rebuilt T19 763 with higher pitched Belpaire boiler, withdrawn in 1913 before conversion to 4-4-0. (Loco Publishing Co./MLS Collection)

Rebuilt T19 702 on a Liverpool Street–Cambridge stopping train at Waltham Cross, 20 June 1910. It was withdrawn in 1919 still as a 2-4-0. (G.M. Shoults/MLS Collection)

In 1900 the 'Claud Hamilton' 4-4-0s appeared (LNER D14) able to take more substantial loads than the 2-4-0s and, as noted above, some 2-4-0s were equipped with similar Belpaire boilers, but in January 1905 a more significant rebuilding of the T19s was put in hand, converting 1035 to the 4-4-0 wheel arrangement followed by fifty-nine more, completed by 1908. They were classified as T19Rebuilt (RBT), all but eight being superheated from 1913. The key dimensions of the 4-4-0 rebuilds were:

Cylinders (2 inside):	18 x 24in
Coupled wheel diameter:	7ft 0in
Bogie wheel diameter:	3ft 1in

Stephenson motion with slide valves

Boiler pressure:	180lbs psi
Heating surface:	1,452.7sq ft
Grate area:	21.6sq ft
Weight (Engine):	48 tons 6 cwt
(Tender):	30 tons 13 cwt (Watercart ex oil: 32 tons 9 cwt)
(Total):	78 tons 19 cwt
Axleload:	17 tons 5 cwt
Water capacity:	2,640 gallons (Watercart 2,790 gallons)
Coal capacity:	5 tons
Tractive effort:	14,163lbs

The sixty rebuilds were:

1905:	700, 706, 707, 718, 729, 748, 765, 775, 1033, 1035
1906:	704, 708, 712, 715, 728, 745, 772, 777, 779, 1012, 1013, 1015, 1016, 1021, 1023, 1025-1027, 1030, 1032
1907:	705, 713, 717, 730, 731, 734, 735, 737, 739, 741, 742, 744, 751, 1018, 1020, 1028, 1029, 1036, 1039
1908:	710, 732, 733, 738, 747, 756, 766, 767, 1031, 1037

The frames were extended at the front end by 2ft 5½in and bogies were retrieved from withdrawn

The newly rebuilt 1035 in its special light grey livery in January 1905.
(F. Moore/MLS Collection)

0-4-4Ts and Worsdell 4-4-0s. The small diameter bogie wheels were a weakness in the design and had a tendency to run hot on the faster services. The first rebuild, 1035, was painted a light grey and was known by its crews as *Dolly Grey* although this unofficial name was never carried. Superheaters of the Schmidt type were first fitted in 1913. After 1915, Robinson superheaters were used (see Appendix for details). The superheated engines could be recognised as they had longer smokeboxes. They were also equipped with mechanical lubricators. At the Grouping thirteen of the rebuilds had 12 ½ in throw crank axles with piston stroke of 25in - 704 - 706, 708, 712, 728, 731 - 733, 742, 765, 772, 777, 779, 1012, 1032 and 1036.

The LNER classified the T19 RBTs as D13 and added 7,000 to the numbers of all ex GE locomotives, so they ran from 7700–7799 and 8012–8039. Two T19 RBTs had been withdrawn by the GE – 715 and 747. All had been fitted with Westinghouse brakes, but 760-764 and 1030–1039 had been also given vacuum ejectors from new and the LNER equipped the rest during 1928/9. The D13s were repainted from the GER dull grey to lined apple green from 1923 until like other LNER 4-4-0s, they reverted to black in 1928. 730 and 1018 were withdrawn before receiving their LNER numbers, and four more were withdrawn in the 1920s, but the majority were withdrawn in the 1930s, the last survivors being 8035 (5/1943), 8023 (1/1944) and 8039 (3/1944).

1025 rebuilt from T19 in 1906, superheated in 1914, seen here with former oil-burning 'Watercart' tender, c1920. It was renumbered 8025 at the Grouping and withdrawn in 1937.
(MLS Collection)

166 • LONDON & NORTH EASTERN RAILWAY 4-4-0 TENDER LOCOMOTIVES

7751, rebuilt from T19 751 in 1907, and superheated in 1914, with step shaped frame in front of the smokebox, with a tender in LNER lined apple green livery, c1927. (MLS Collection)

8026, rebuilt from 1895 constructed T19 1026, out-shopped in May 1926 newly superheated by the LNER, with a 'watercart' tender, in lined apple green livery. (F. Moore/MLS Collection)

8020, ex-works in black livery at Stratford, 1932. It was rebuilt as a D13, 1020, in 1907, superheated in 1914 and is coupled with a 'watercart' tender. Behind is F3 2-4-2T 8044. (J.A.G. Coltas/MLS Collection)

7707, the second T19 2-4-0 to be rebuilt as D13 707 in January 1905, superheated in 1913, seen here in its last year before withdrawal, c1936. It has the stepped frame in front of the smokebox. (MLS Collection)

1035, the first T19 to be rebuilt in January 1905, seen here running in on a stopping train in its light grey livery, and with the original smokebox door shape. 1035's livery was much lighter than the later GE dull grey, and this engine was known affectionately by its crews as *Dolly Grey*. (E. Pouteau/MLS Collection)

Operation

The T19s worked the main Great Eastern expresses along with the Holden 'D7' 2-2-2s of 1889 until the debut of the 'Claud Hamiltons' in 1900. By the time of their rebuilding as 4-4-0s, the prime express work, apart from the Liverpool Street–Cambridge line was firmly in the hands of the 'Clauds' and then, shortly afterwards, the 1500s (the 'B12s'). The T19 RBTs were therefore allocated over the whole Great Eastern Railway system for semi-fast, slow passenger and branch trains. They lost some of their Cambridge work as more 1500s became available releasing 'D14s' and 'D15s'. The allocation of the remaining fifty-eight D13s, as they were now classified, was at the beginning of 1923:

Stratford:	3
Colchester:	7
Parkeston:	1
Ipswich:	5
Norwich:	4
Yarmouth:	2
Cambridge:	12
King's Lynn:	3
March:	6
Peterborough East	8
Lincoln:	3
Doncaster:	4

Most had departed from the London depots by 1930, although occasionally one would be found on a slow Southend train. Peter Proud, a regular commuter from Broxbourne to Liverpool Street, occasionally found a D13 on his regular services between 1931 and 1933 before D15s or D16s took over. Noted were 7779 and 8012 in particular. The last to leave was 8039 which had a good reputation and could often be seen as station pilot at Liverpool Street ready to cover for any failures. B17s started to come on stream from 1929 and the cascade of power resulted in more D15s and D16s being available

730 built in 1888 and rebuilt as a 4-4-0 in 1907, piloting an S46 (LNER D14) on a Down Norwich express, c1912. It has the original smokebox door and sports fixed lamps on the centre of the buffer beam and on the smokebox but carries discs in the GE tradition to show the express headcode. (F. Moore/MLS Collection)

for the D13 turns. Five of the eight unsuperheated D13s had been withdrawn in the 1920s, followed by two more and a superheated example in 1929/30, leaving fifty in service at the beginning of 1932. They worked some stopping trains from Cambridge and Bishop's Stortford although they were being replaced on those duties by N7 0-6-2Ts and Gresley V1 2-6-2Ts from the mid-1930s. With their Belpaire fireboxes, they could produce short bursts of power to give reasonable acceleration for a 4-4-0 and could be speedy when given the opportunity, although their riding could be exciting! Their main work was concentrated around Cambridge and its sub-depots and they frequently found use on horsebox specials from the Newmarket area.

The March–Lincoln area through the Fens was another area where they lingered. Where there was significant tender-first working, the ex-oil 'watercart' tenders were preferred as sighting was better.

Peter Proud was a regular commuter between Broxbourne and Liverpool Street in the 1930s and timed every trip, the Railway Performance Society holding pages of his records. Some of the Cambridge engines sub-shedded at Bishop's Stortford featured heavily, especially in the evening down direction. 7712, 7732, 8021, 8023, 8027 and 8028 were noted on occasions, but 7779, 8012 and 8016 spent weeks, day after day, between 1932 and 1935 on the 7.35am or 10.24am Broxbourne in the morning and the 12.35pm or 2.01pm Down,

or one of the evening commuter trains. The morning Up trains weighed 220-290 tons, the early afternoon trains were lighter, 115-120 tons, but the evening commuter trains were heavier, up to 310 tons. Some trains ran non-stop to Broxbourne, others stopped at Waltham Cross. The 'fast' trains generally clocked a maximum speed of around 50–55mph, the lighter midday trains 56-59mph. I cannot trace any runs where 60mph was reached. The heavier trains stopping at Tottenham Hale and Waltham Cross barely made 50mph between stops. After the steep climb out of Liverpool Street to Bethnal Green, the Down run was against the grain with patches of level track alternating with a mile of 1 in 210 up before

Ponders End and four miles of 1 in 731 up between Waltham Cross and Broxbourne. The schedule for non-stop trains over the 17 miles was 27 minutes. On a couple of occasions 7779 on the heavy 7.35am Up struggled to exceed 45mph after the Waltham Cross stop, with comments that there was much slipping. By 1935, D15s and D16s had largely taken over the commuter trains.

Withdrawals increased rapidly during the 1930s and in 1935 just twenty-nine were left, based at:

Colchester:	7756
Ipswich:	7744, 8025
Norwich:	7766, 8013, 8026, 8027, 8032
Yarmouth Vauxhall:	7702, 7772
Cambridge:	7779, 8012, 8016, 8021, 8023, 8038
King's Lynn:	7700, 7708, 7729, 7741, 7742, 8029, 8036
March:	7706, 8020
Peterborough East:	7775, 8030, 8035, 8039

As well as being replaced by D15s and D16s, a few ex-GC D9s appeared in East Anglia and at the beginning of 1939, the last three D13s were based at Cambridge (8023 and 8039) and Norwich (8035). 8035 returned to King's Lynn at the end of 1939 but joined the other two at Cambridge in October 1940. Their last general repairs were in 1940 and the war demands kept them active. They were averaging around 25,000 miles a year until 8035 was withdrawn in 1943 and the last two in January (8023) and March 1944 (8039). 8039 had run 78,303 miles since its last major repair, a good performance for the class.

The GE S46 (LNER D14)

Although the design was ascribed to James Holden, in fact most of the design work of this 4-4-0 class was undertaken by his Chief Draughtsman, Frederick Russell, as Holden was in poor health at the time. This was the Great Eastern's first 4-4-0, the prototype, No.1900, built in 1900, and named *Claud Hamilton* after the company's Director, some five years before the rebuilding of the T19 2-4-0s as 4-4-0s. (Claud Hamilton was a GER Director from 1872 and Chairman from 1893–December 1922.)

A further ten, numbering backwards, 1890–1899 from the prototype, were completed in 1900 also and were followed by a further ten, 1880–1889 in 1901, another

8032 built in 1897, rebuilt in 1906 and superheated in 1921, at Dereham, with J15 7904, 8 July 1936. 8025 was another D13 with 25in stroke and 'watercart' tender. It was withdrawn just two months later. (Pamlin Prints/MLS Collection)

ten, 1870–1879 in 1902, and ten more, 1860–1869, in 1903. Their key dimensions were:

Cylinders
  (2 inside):   19 x 26in
Coupled wheel
  diameter: 7ft 0in
Bogie wheel
  diameter: 3ft 9in

Stephenson motion and slide valves
Boiler Pressure
  (round top):   180lbs psi
Heating surface: 1,630.5sq ft
Grate area: 21.3sq ft
Axleload: 16 tons 12 cwt

Weight (Engine):   50 tons 6 cwt
       (Tender):   39 tons 5 cwt
       (Total):   89 tons 11 cwt
Water capacity: 2,790 gallons
Oil capacity: 720 gallons
                       + 1½ tons
                       of coal for
                       lighting up.
Coal capacity
  (post 1904):   5 tons
Tractive effort: 17,096lbs
Westinghouse air brake

1900 was exhibited in 1900 at the Paris Exhibition and was awarded a Gold Medal. 1900 and the following forty S46s were built as oil-burners. The following D56 class (LNER D15s) were built as coal burners and subsequently from 1911 the GE abandoned oil-burning. Three were, however, refitted for oil-burning in 1916 (1860, 1862 and 1864) and during the 1921 coal strike, forty-four were re-equipped and the same for a few months during the general strike of 1926. The engines were fitted with water scoops as water troughs had been constructed at Ipswich and Tivetshall to allow non-stop running to Norwich or North Walsham a few years previously. The cab width was increased from 6ft 3in to 7ft 2in from No.1880 and had a high arched roof. A larger

1900 *Claud Hamilton* in exhibition finish on display at St. Pancras station prior to shipping to France for the March 1900 Paris Exhibition. Although an oil-burner with 'watercart' tender, the coal used for lighting up can be seen on the tender. (MLS Collection)

capacity tender was provided for No.1870 onwards holding 3,300 gallons of water and the earlier engines were similarly equipped later. New round-topped boilers with 1,624.38sq ft of heating surface and grate areas of 21.6sq ft were provided to all the S46s between 1909 and 1913, apart from 1866 and 1868 which received secondhand boilers in 1916 and 1915 respectively.

The livery of GE passenger engines from 1882 had been royal blue and this was used for all the 'Claud' 4-4-0s with few exceptions – a couple, 1813 and 1855 were grey and 1830 black. 1885 became the GE's royal train engine and was given additional decoration and lining which was added to a number of other locomotives subsequently. During the First World War, the blue was replaced by a matt grey unlined livery with GER on the tender until 1921 when the engine number in large yellow figures was painted on the tender sides. In 1923 the LNER lined apple green became the livery for all the 'Claud' variations, although a number of engines for a few months emerged from Stratford Works still in matt grey with the LNER number on the tender. In 1928 the LNER adopted black with red lining for all but front line express locomotives and running numbers transferred to the cabside. Although the former brass beading over the splashers was painted over at overhauls, many shed cleaners removed the paint to reveal the beading to set off the black livery. The rebuilt D14s (and later D15s) were painted plain black after nationalisation with BR number and 'lion and wheel' totem on the tender.

1896 and 1898 were rebuilt with superheated Belpaire boilers as class D56 (LNER D15) in 1915 and twenty-six more between then and 1927, and eight with saturated Belpaire boilers between 1921 and 1930, the remaining unrebuilt S46s, D14, from the Grouping, were withdrawn by 1931. Twenty-eight

1872 as built in 1903 but with 3,300 gallon tender after the cessation of oil-burning, seen here c1910. The smokebox door has a reinforcing steel ring, which became the pattern for all the 'Clauds', often kept burnished by cleaners at the local depot, as here. (F. Moore/MLS Collection)

8879 in LNER lined green livery at Stratford, c1926. 8879 was rebuilt as a D15/2 in 1927. (MLS Collection)

of the D14s that had been converted to D15s were ultimately rebuilt as D16s.

The GE D56 (LNER D15)

The Belpaire firebox first appeared in the UK in 1891 and by the turn of the century it was being used on the locomotives of the Great Central, Lancashire & Yorkshire and Great Western Railways. Holden sought to equip a 'Claud Hamilton' with one in 1901, but problems were encountered in the construction and an 0-6-0 was the first to receive one on the Great Eastern. The problems had been sorted by 1903 and the next batch of ten 4-4-0s in the S46 sequence, 1850 – 1859, were turned out with Belpaire boilers and the GE locomotive department reclassified them as class D56.

The overall size of the boiler was similar, but the main differences in the dimensions were:

Heating surface: 1,706.58sq ft
Weight (Engine): 51 tons 12 cwt
(Total): 90 tons 17 cwt

Deeper frames were provided and the boiler was pitched 2in higher and therefore a shorter chimney and dome and shallower cab roof were needed to meet the GE's restricted loading gauge. 1850–1859 were the last 'Clauds' built as oil burners with the 'watercart' tender. Sixty more, 1790–1849, were constructed between 1906 and 1911. Four of the 1911 series, 1793, 1794, 1798 and 1799, were equipped with superheated boilers, the first two with the Schmidt type and the latter two with the GW Swindon pattern. The revised heating surface of these four engines were:

1793 & 1794: Heating surface 1,501.1sq ft (including 226sq ft of superheater surface)
1798 & 1799: Heating surface 1,550.8sq ft (including 188sq ft of superheater surface)

The Robinson superheater was later adopted as standard, tested initially in 1914 on 1792, 1797, 1807 and 1813. The heating surface was identical to the engines with the Schmidt superheater but the engine weight increased to 54 tons 9 cwt and the maximum axleload to 18 tons 1 cwt. A revision was made to the Robinson superheater design in 1921 using short return

D56 1828 as built in December 1909 in GER royal blue livery. (Real Photographs/MLS Collection)

loop superheater elements reducing the heating surface from 226 to 154.8sq ft and the total heating surface to 1.429.9sq ft. The sixty were built as coal burners and the last pair of each batch of ten were equipped with vacuum as well as Westinghouse air brakes.

At the Grouping, twenty of the earlier S46 (D14) engines had been rebuilt with Belpaire boilers and with the 1903-1911 D56s, made a class of 90 D15s. The LNER lined apple green livery was applied to the Clauds from April 1923, although still with their GE numbers. The suffix 'E' was added to the number of a new build of ten 'Clauds' between June and September (1780-1789) and about twenty others before 7000 was added to the numbers of former GE engines from February 1924. The D15s therefore became 8790-8900 inclusive of the rebuilt and unrebuilt D14s. The twenty-one remaining D14s were rebuilt as D15s with the Belpaire boiler between 1921 and 1931. The rebuilt D14s were:

1915:	1896, 1898 (superheated)
1916:	1883, 1890, 1891 (superheated)
1917:	1889 (superheated)
1918:	1880, 1887, 1888, 1894 (superheated)
1919:	1860, 1874, 1882, 1885, 1897 (superheated)
1921:	1861, 1863 (saturated)
1922:	1869, 1884, 1899 (saturated)
1923:	1862, 1864, 1866, 1867 (superheated)
1924:	8870-8872 (superheated)
1925:	8900 (superheated), 8886 (saturated)
1926:	8893, 8895 (superheated)
1927:	8879, 8892 (superheated)
1928:	8877 (superheated)
1929:	8868, 8876 (superheated), 8878 (saturated)
1930:	8865, 8881 (superheated), 8873 (saturated)
1931:	8875 (superheated)

In 1926, superheated 8819 was fitted with an extended smokebox and this became the standard fitting for the D15s with superheated boilers. In 1927, the LNER reclassified the D15s with the extended smokeboxes as D15/2, those retaining the short

8900 *Claud Hamilton* newly rebuilt with a superheated boiler at the Darlington 1925 Centenary Exhibition. (Real Photographs/MLS Collection)

1791 built in 1911 as a D56, seen here around 1916 at Stratford in grey livery, fitted with feed water heater apparatus before being superheated in 1923, and rebuilt with an extended smokebox boiler in 1928. (MLS Collection)

smokebox, whether still saturated or superheated, as D15/1s. The last saturated D15/1 with short smokebox was 8873, rebuilt as a D15/2 in 1933 and the last superheated D15/1 to be rebuilt as a D15/2 was 8812 in April 1934, although 8899 which had got an extended smokebox boiler in 1929 lost it during a Works visit in 1933 and only reverted to an extended smokebox boiler in March 1935.

From 1928 the livery was changed to black with red lining. The lining was omitted from November 1940 because of wartime conditions, with plain 'NE' on the tender. From 1946, the full 'LNER' replaced this and the two former royal Clauds, 8783 and 8787 were repainted lined

8828, a D15 still with a saturated short smokebox boiler, in LNER lined green livery at Stratford, c1924. (MLS Collection)

green as 2614 and 2618. The D15s were all painted plain black in BR days unlike the D16s which received the mixed traffic lined black – apart from D15 62538 which, after being painted plain black during a Works overhaul in 1950, was returned by its depot for lining out at the insistence of its regular crew! A problem that developed with the D15s during the latter half of the 1920s was both unsatisfactory steaming and heavy coal consumption. The small coal was getting crushed in the new mechanical coaling plants constructed by the LNER and it was working forward in the steeply sloping grates blocking the air flow. Initial experiments to tackle the problem were centred around the blastpipe and chimney design and a GN style chimney was fitted to a number from 1930. The sloping of the grate was realised to be part of the problem and from 1934 the angularity of the grate base was reduced and was found to give a significant improvement in both steaming and coal consumption.

The vast majority of D15s were rebuilt as D16s from the mid-1930s (see next section), but thirteen remained as D15/2s and were withdrawn between 1948 and 1952. These included nine of the earliest S46 batch built in 1900, 1890–1898, renumbered 2501–2509 in 1946 and 62501–62509 in 1948. The last survivor was 62509, withdrawn in September 1952.

D15/2 2505, formerly 8894 seen here in 1946. (MLS Collection)

D15/2 2502, renumbered from 8891 in 1946, at Yarmouth South Town, c1949. It is still coupled to a former oil-burner 'watercart' tender. (MLS Collection)

D15/2 2502 again, still with its watercart tender, but a couple of years later renumbered 62502. It is seen at King's Lynn, May 1951. It was withdrawn in May 1952. (J. Davenport/MLS Collection)

D15/2 62506 also photographed at King's Lynn in May 1952. 62506, the former 8895, became a superheated D15 in 1926, classified D15/2 in 1930 when it was fitted with a boiler that had an extended smokebox. (J. Davenport/MLS Collection)

The GE H88 (LNER D16)

After the construction of 1790–1799 in 1911, a new 4-6-0, the 1500 class, (LNER B12) was designed and constructed as the Great Eastern's main express passenger locomotive. The 4-6-0 had a larger superheated boiler and longer firebox and a version of this boiler, only 9 inches shorter than the 4-6-0s and fitted with a 21-element superheater, was fitted to D56 1805 in early 1923, still in the post-war GE dull grey livery. It became known as a 'Super Claud' and was given the classification of D16 by the LNER motive power department. Ten new engines with this boiler were constructed in 1923 using the GE class identification of H88, although they were soon designated by the LNER, with 1805, as D16s. The new engines were numbered 1780E–1789E and emerged in the lined green livery.

The main dimensional differences to the preceding D15s were:

Boiler diameter:	5ft 1½in (instead of 4ft 9in)
Boiler pitch:	68ft 8in (instead of 8ft 5in)
Heating surface:	1,688.4sq ft (raised from 1429.4sq ft)
Superheater:	21 instead of 18 elements

Ross pop safety valves
Weight (Engine): 54 tons 18 cwt
(Total): 94 tons 3 cwt

The following D15s were rebuilt with the new boilers as D16s to join 1805 and 1780-1789: 1818 in 1923, 8846 in 1924, 8847 and 8853 in 1926. Then in April 1926, 8813 was similarly rebuilt but was given an extended smokebox and was given the classification of D16/2. The 1923 built engines and the five rebuilds with the short smokebox were reclassified as D16/1. All subsequent rebuilds were to the D16/2 class. They were:

1927:	8800, 8827, 8843, 8851, 8856
1928:	8792, 8796, 8819, 8831, 8839, 8852
1929:	8790, 8794, 8801, 8826, 8834, 8838, 8841, 8845
1930:	8822, 8829, 8833, 8842
1931:	8795

The fifteen D16/1s were all given extended smokeboxes between 1928 and 1934, converting them to class D16/2. 8781, as the result of collision damage in 1931, had a changed shape to the footplating forward

GE H88 (LNER D16) 1785E as built in July 1923 and finished in LNER lined apple green livery but retaining GE numbering with the addition of the suffix 'E'. (MLS Collection).

D16/2 8839, one of six D15s rebuilt with extended smokebox in 1928 at Stratford, April 1928. (MLS Collection)

D16/2 8781 which was badly damaged in a collision at Thorpe-le-Soken in January 1931, had reshaped frames at the front end as seen here in this photograph taken c1932. This altered shape is evident on D16s that had frame repairs subsequently. (MLS Collection).

D16/2 8787, one of the two regular royal engines, after extension of smokebox, January 1929. (MLS Collection)

8796 in LNER black livery, converted from D15 to D16/2 in April 1928, seen here c1930. (MLS Collection)

of the smokebox, a shape that was adopted for later rebuilds (see D16/3 later). At first, the 'Super Clauds' faced the same steaming difficulties the D15s and the change to the grate slope to 8900 and some D15s does not seem to have been applied to the 'Super Clauds' in the short term.

Edward Thompson, as Mechanical Engineer at Stratford, had overseen the fitting of a B12 with long travel valves and an enlarged round-topped boiler in 1932 and had fitted a new Claud round-topped boiler (type 28A) to D15/2 8848 in January 1933. 8848 was re-designated class D16/3 and had the following changed dimensions:

Boiler pitch:	8ft 9in
Heating surface:	1,633.8sq ft (incl superheater 204.4sq ft)
Grate area:	21sq ft
Axleload:	18 tons 14 cwt
Weight (Engine):	55 tons 18 cwt
(Total):	95 tons 3 cwt

Thompson had sought approval to include new cylinders with piston valves with long travel valve gear in the rebuilding following the success of the rebuilding of some of the B12s, but this was initially turned down by Gresley. In 1933, the proposal was again submitted and agreed provided standard parts as used in the J39 0-6-0 were used. The second rebuild, 8900 *Claud Hamilton*, was given new cylinders with piston valves at the same time as receiving the 28A boiler and nine further engines were provided with new cylinders and piston valves when rebuilt as D16/3s in the following eighteen months. They were 8798, 8804, 8809, 8816, 8817, 8837, 8849, 8855, 8866. At the same time, other 'Clauds' were rebuilt with the round-topped boiler but retaining slide valves. When the piston valve engines demonstrated their superiority in service, a decision was made in 1936 to fit a number of the D16/3 conversions that had retained slide valves with larger 9½ in piston valves – 8791, 8803, 8808, 8810, 8812, 8825, 8832, 8864 and 8865. This, however, proved to have one disadvantage. The greater power produced by these engines put extra stress on the frames and fractures occurred at the weakest point at the rear of the bogie wheels. Various repairs including patch welding were tried, but the problem was never satisfactorily resolved and as a result some of the locomotives with piston valves were withdrawn prematurely because of severely cracked frames. These frame problems had not been corrected by the onset of the Second World War and all subsequent conversions to D16/3 retained slide valves.

The rebuilding of D15s to D16/3 included some of the original S46 series (D14s) as the 28A boiler could be fitted on their frames – the differences in these frames caused the rebuilds initially to be designated as class D14/2, which consisted of 8860, 8861, 8863, 8866, 8869, 8870, 8876, 8878 and 8900. However, in 1936 the different designation was thought to be immaterial, and all rebuilds with round-topped boilers were thereafter known as D16/3s.

The equipping of 8900 with a lessened sloping grate to cure the poor steaming problems encountered by the D15s was at length considered relevant for the D16/3 rebuilds which had also shown similar steaming difficulties. The length of the grate had been slightly shortened from 6ft 5in to 6ft 3¼ in and the area from 21.4sq ft to 21.0sq ft. The rebuilding of D16/2s as well as D15s and D16/1s to D16/3s with the Gresley/Thompson round-topped boiler followed rapidly. They retained the decorative valances below the running plate which had been removed from the D15s which had been rebuilt as D16/3s. In all, 104 of the 121 'Clauds' were eventually rebuilt as class D16/3s. Thirteen D15/2s had remained until withdrawal as stated earlier and just four D16/2s with Belpaire boilers remained unconverted until their withdrawal between 1950 and 1952 – 62547, 62590, 62591and 62603.

The conversion dates were:

1934:	8802, 8809*, 8821, 8837*, 8848, 8849*, 8854, 8866*, 8900*
1935:	8798*, 8804*. 8816*, 8817*, 8828, 8855*, 8859, 8860, 8869
1935:	8814, 8840, 8863, 8870, 8876, 8878
1936:	8810*, 8812*, 8832*, 8861*, 8864*, 8865*
1937:	8791*, 8793, 8797, 8803*, 8808*, 8825*, 8836, 8875, 8885
1938:	8788, 8823, 8843, 8851, 8858, 8871, 8874, 8879
1939:	8783, 8811, 8815, 8820, 8835, 8844, 8850, 8857, 8873, 8888
1940:	8794, 8799, 8806, 8824, 8830, 8862, 8868, 8872, 8886
1941:	-
1942:	8807, 8880, 8882
1943:	8883, 8884, 8887, 8899

1944: 8780, 8785, 8787, 8790, 8827, 8846
1945: 8786, 8801, 8834
1946: 2607, 8795, 8831, 8845
1947: 2544, 2573, 2584, 2589, 2596, 2615
1948: 2564, 2620, 62558, 62569, 62580, 62613
1949: 62543, 62552, 62553, 62570, 62577, 62612

* piston valve engines

The first withdrawals of the D16/3s occurred while some rebuildings were still taking place. This occurred because of the frame fractures of some of the piston valve engines as stated earlier. The first to go was 8866 in September 1945. Others withdrawn before nationalisation were:

1945: 8866*
1946: 2550, 2595*
1948: 2560*, 2563*, 2583*, 2600*, 2602*

The final survivors, which lasted to 1960 despite the rapid dieselisation of East Anglia from 1958, were 62524, 62597, 62604 and 62613, the latter being the very last withdrawn in October.

D14, D15 & D16 – Operation

The S46s took over the main Great Eastern services to Ipswich and Norwich, Parkeston Quay and Clacton from the T19 2-4-0s which had not at that stage been rebuilt as 4-4-0s. As numbers grew, they also worked the most important expresses on the Southend line and services through March to the north. The most prestigious service they worked and where their early work was recorded was on 'The Norfolk Coast Express', which ran between Liverpool Street and Cromer, non-stop as far as North Walsham, just over 130 miles. Cecil J. Allen published a run each way on this train in the November 1911 edition of the *Railway Magazine*, with heavy trains hauled by the latest D56 (D15) locomotives.

The pioneer engine, 8900 *Claud Hamilton* at Startford after rebuilding as a D16/3 with 8in piston valves and round-topped boiler, February 1933. (Rail Archive Stephenson/John Scott-Morgan)

8812 rebuilt as a D16/3 in 1936 with round-topped boiler and 9½in piston valves, at March, 10 June 1938. (MLS Collection)

D16/3 2548 at March, 25 April 1948. It was superheated in 1916, rebuilt as a D15/2 (as 8857) in 1933 and as a D16/3 with slide valves in July 1939. It has a GN style chimney, the decorative valance above the coupled wheels has been removed and the engine received its revised number in June 1948. It was withdrawn in October 1957.
(J. Davenport/MLS Collection).

62521 rebuilt from D14 to D15 in 1924 and converted to a D16/3 with slide valves in 1935. It was withdrawn in February 1958. It is seen in the company of F5 2-4-2T 67154 at Yarmouth South Town, 19 May 1951. (MLS Collection)

62546 named *Claud Hamilton* when 8900 was withdrawn in May 1947, seen here at Yarmouth South Town in BR mixed traffic lined black, 26 May 1956. (MLS Collection)

Ex-works 62530 at March, 4 May 1958. (MLS Collection)

One of the two D16/3s painted apple green after the Second World War and used for royal train working, 62618, at King's Lynn, c1956. (MLS Collection)

In 1910, a S46 (D14) with a 15 coach 390 ton train ran the 68 ½ miles to Ipswich in 84½ minutes, dropping just half a minute on the schedule, recovering after a painful climb to Ingrave summit started at just 29mph and completed at 18 and passing Shenfield four minutes down. Another unidentified D14 had twelve coaches, 350 tons, on the up *Norfolk Coast Express* in January 1916 and held the schedule meticulously between a quarter and one minute early all the way from Wroxham to Brentwood, but signal checks from Romford onwards made the train ten minutes late at Liverpool Street.

The first 1500 4-6-0s appeared in 1911 and displaced the 'Clauds' from the fastest and heaviest turns and the 4-4-0s then began to appear more frequently on the Cambridge and King's Lynn route., where they in turn replaced the rebuilt T19s (D13s). The D15s retained their express work on the Clacton route until the track was upgraded in 1930, although with the increase in train weights, some double-heading by D15s and D16s took place in the mid-late 1920s. At the Grouping in 1923 the allocation of the S46s (D14s) and 'D56s' (D15s) was as follows:

Stratford: 35 (including 7 of the D16 'Super Clauds')
Colchester: 6 (including 2 of the D16 'Super Clauds')
Ipswich: 19 (including 1 of the D16 'Super Clauds')
Parkeston: 3
Norwich: 15
Yarmouth: 5
Lowestoft: 1
Cambridge: 19
King's Lynn: 2
March: 5
Doncaster: 1

62618 again at March on 10 August 1958 alongside another 'Claud' and in front of a LNWR 0-8-0. It has lost its green livery (the 'royals' are now catered for on the GE section by 'Britannias' and one of the two B2s used for the purpose). 62618 is now in BR mixed traffic lined black. *(MLS Collection)*

		The 'Norfolk Coast Express', 1911						
		1823 (built 1909)			1809 (built 1910)			
		12 chs, 330 tons			14 chs, 400 tons			
		Driver Storey			Driver Cage			
		Down run			Up run			
Miles	Location	Times	Speed		Times	Speed		Gradients (Down)
0	Liverpool Street	00.00		T	157.24		1 ½ E	
1.3	Bethnal Green	03.21	18		155.10	38		1/70 R
3.7	Stratford	07.58	37	1 E	150.40	sigs/57	¼ E	1/350 F
9.7	Chadwell Heath	15.53	51	¼ E	144.20	76 ½	¼ L	1/390 R
19.2	Ingrave summit	29.09	26 ½		135.39	36		1/103 R
	Ingatestone	-	71 ½		-	46		1/135 F
29.6	Chelmsford	39.28	30*	½ L	121.46	53	1¼ E	
38.4	Witham	48.43	57	¼ E	111.51	56	1 E	1/181 R, 1/178 F
46.5	Marks Tey	56.39	66		103.05	51		L, 1/222 R
51.5	Colchester	61.26	61 ½	2½ E	97.20	52 ½	¾ E	
	Ardleigh	-	45		-	40 ½		1/144 R
59.2	Manningtree	70.15	65	2¾ E	88.29	67	½ L	1/134 F
65.1	Belstead	77.02	40 ½		81.39	31		1/157 R, 1/130 F
68.5	Ipswich	80.51	65/25*	2¼ E	76.25	66/25*	½ L	1/120 F, L
80.6	Stowmarket	95.09	56 ½	1¾ E	65.25	70 ½	1½ L	L
82.7	Haughley	97.41	40	2¼ E	63.30	47 ½	1½ L	1/131 R
86.5	Finningham	102.41			59.14	51 ½		1/248 F
95	Diss	110.41	75		49.25	66/55 ½		L, 1/132 F
100.4	Tivetshall	116.14	59/69	3¾ E	43.31	42 ½	½ L	L, 1/138 F
112.6	Trowse Upper Jn	128.31	70/ 15*		26.08	15*		1/135 F, 1/134 R
114.6	Wensum Junction	133.20	15*	3¾ E	21.35	42 ½/15*	½ L	L
122.8	Wroxham	147.33	34 ½/sigs		10.04	42/35*		
130	North Walsham	160.00	sigs (153 net)	2 L	00.00		1 L	

The maximum effort by the D16s has probably not been recorded because by the time of their superheating and rebuilding, the 1500s were in charge of the expresses requiring the maximum effort. However, four runs between Ipswich and Norwich were published in Cecil J. Allen's column in the November 1937 edition of the *Railway Magazine*. Unfortunately, he does not quote the dates of the runs so although the number of the locomotive is given, it is not known if this was a D14 or D15 in original condition or if after superheating or rebuilding. I suspect they were in Great Eastern days when the 'Clauds' were still on top link work. However, the number quoted is the LNER post 1924 number, so they

D56 (later D15) 1816 heads the Up *Norfolk Coast Express* near Shenfield, 1 September 1911. (G.M. Shoults/MLS Collection)

D56 1842 with the Norfolk Coast portion of an express for Liverpool Street, c1910. (Loco Publishing Co./MLS Collection)

190 • LONDON & NORTH EASTERN RAILWAY 4-4-0 TENDER LOCOMOTIVES

D56 1858 with a Southend train near Shenfield, c1920
(MLS Collection)

may be runs from the 1920s. The three locomotives quoted have the following history and the runs could have been in any of these states:

- 8808: built in 1910 as a D56 (LNER D15), D15/2 with extended smokebox in 1929, D16/3 with large piston valves in July 1937.
- 8870: built in 1902 as an S46 (LNER D14), rebuilt as a superheated D15 in 1924 and with extended smokebox in 1928., D16/3 with slide valves in 1935.
- 8864: built in 1903 as an S46, rebuilt as a superheated D15 in 1923, and D16/3 with large piston valves in 1936.

Ipswich–Norwich, unknown date

		8808 8 chs, 260/280 tons		8870 9 chs, 293/310 tons		8864 9 chs, 231/245 tons 5.16pm L.St – Cromer			
Miles	Location	Times	Speed	Times	Speed		Times	Speed	Gradients
0	Ipswich	00.00		00.00			00.00		
2.5	Bramford	04.40		04.45			04.15		L
4.9	Claydon	07.25	56	07.45	58		06.51	61	L
8.4	Needham	11.15	63	11.30	63		10.15	63	1/337 R
11.9	Stowmarket	15.10	¾ E	15.20	¾ E		13.30	65	
0		00.00	T	00.00	T				1/148 R
2.3	Haughley	04.30	40	04.35	39		15.50	52	1/131 R
6	Finningham	09.20		09.37			19.45		1/248 F
10.8	Mellis	13.37	83	14.10	82		23.50	86	1/132 F
14.4	Diss	16.23	82	17.02	75		26.30	81	L, 1/558 R
19.9	Tivetshall	21.08	65 ¼ L	22.00	61	1 L	31.00	65	1/142 R
23.5	Forncett	24.04	80	25.10	79		33.58	83	1/138 F, 1/136 F
26.1	Flordon	26.01	76	27.14	66		35.55	70	1/134 R
29.1	Swainsthorpe	28.20	84	29.46	78		38.22	78	1/143 F, L
31.4	MP 112	30.05	82	31.35	76		40.12	77	
33.4	Trowse Upper Jn	32.43		34.05			42.45	sig stop – 7 mins	
34.4	Norwich	34.37	2½ E	36.00	1 E		51.45	(44¾ net) 1¾ L	

The Great Eastern borrowed a GNR Pullman set on Summer Sundays in 1922 to run *The Clacton Pullman* and when the LNER constructed new Pullman cars for East Coast services in 1929, the redundant sets were released for weekday use in East Anglia and formed 'The Eastern Belle Pullman Limited' until the outbreak of the Second World War. This ran to various locations – Clacton, Felixstowe, Yarmouth, Sheringham and Cromer, Hunstanton and even Skegness – at excursion fares. It was formed of seven Pullmans and in the first few years was normally hauled by a 'Claud', although B12s and B17s became more common in the late 1930s. The 'Clauds' were also the staple power for Newmarket race specials, starting from King's Cross.

A significant change in allocations occurred on the delivery of the B17 'Sandringham' 4-6-0s in 1929 and 1930 and the removal of the weight restrictions on the Clacton line. By this time, forty of the 'Clauds' had been rebuilt as D16s, although they had not been concentrated at any depots but were fairly evenly distributed over East Anglia along with the D15s. The main impact of the B17 arrivals was a reduction of 4-4-0s at Stratford (-9) and an increase at Norwich for working on the branches to the coast.

8900 *Claud Hamilton* after its part in the Darlington Centenary Parade in March 1925, is seen later in the year south of Shenfield, most likely on a Southend train. (MLS Collection)

1923 built D16/1 8786 climbing Brentwood Bank with a Down Norwich express, c1925. (MLS Collection)

D15 8838 with the *Eastern Belle Pullman* excursion train in 1929 shortly before 8838's rebuilding as a D16/2 in November of that year. (MLS Collection)

8873 at Beccles with a southbound express, c1928. 8873 was the last saturated D14 to be rebuilt as a D15 in January 1930. (H. Gordon Tidey/MLS Collection)

8873 rebuilt with an extended Belpaire boiler and classified as a D15/2 in March 1933, at Cromer, c1934. (MLS Collection)

The allocation at the beginning of 1931 was:

		Change from 1923
Stratford:	26 (12 D16s)	-9
Colchester:	10 (4 D16s)	+4
Parkeston:	1	-2
Ipswich:	18 (5 D16s)	-1
Norwich:	20 (6 D16s)	+5
Yarmouth:	6 (1 D16)	+1
Lowestoft:	1	No change
Cambridge:	19 (4 D16s)	No change
King's Lynn:	3 (1 D16)	+1
March:	10 (6 D16s)	+5
Lincoln:	3 (1 D16)	+3
Doncaster:	-	-1
Peterborough	4	+4

In 1932, the LNER introduced five new accelerated expresses each way between Cambridge and King's Cross with stops at Royston, Letchworth and Hitchin. The normal load was just three to five coaches, including a buffet car, and they became nicknamed 'Beer Trains'. They were allowed 75 minutes on the Down run and 72 on the Up. The majority of the turns fell to Cambridge shed to supply power and a number of different classes were used, including D15s and D16s. The locomotives that were eventually associated with these trains until the B17s appeared were the GN Ivatt C1 Atlantics, but the 'Clauds' continued to appear until 1938, especially the two 'royal' D16s, 8783 and 8787.

8810, a piston valve D16/3 was also one of the regular engines used on these 'Beer Trains' and was recorded on 5 November 1936 with seven coaches, 225 tons, taking 26¾ minutes to run the 20½ miles to Welwyn Garden City, touching 80mph after a slow start and no more than 45mph at Potters Bar. 72mph was touched before the Hitchin stop, 11½ miles in 14½ minutes, and 72 again before the Royston stop, barely holding to the schedule. The same engine on 12 July the following year made a better run with the same seven coach load, with 50mph at Potters Bar, 76 at Hatfield reaching

Welwyn G C virtually on time in 25¼ minutes. Welwyn - Hitchin took exactly the scheduled 14 minutes, with 72 maximum and Letchworth–Cambridge was completed ¼ minute early with 76½mph down the 1 in 120 after Royston. An exceptional run was made on 10 August 1936 when 8810 on the 5.25pm 'Beer Train' from Cambridge with just five coaches (155 tons) touched 83½mph at New Barnet and a full 90mph at Wood Green, the highest speed logged by a 'Claud'. A more normal up run was made by 8852 and seven coaches when the 23½ mile Cambridge–Letchworth mainly uphill stretch was completed in 27 minutes 21 seconds, over a minute under schedule with 62 on the level before the two miles of 1 in 120 before Royston where speed dropped to 50mph and recovered to 66½ on the mile descent at 1 in 137 to Baldock. The 6.9 mainly uphill (1 in 200) miles to Knebworth took 9minutes 51 seconds with 56mph at Stevenage and 62 in the dip after and 69 before the Welwyn GC stop. 69mph after Potters Bar would have assured a punctual arrival, but this was spoilt by signal checks approaching King's Cross, reached 2 minutes late. GN Atlantics and B17s were more common on the 'Beer Trains' in the late 1930s, but the two 'royal' engines, 8783 and 8787, continued to frequent these services right up to the outbreak of war.

The piston valve 'Clauds' were equally competent on the 65 minute non-stop runs for the 55½ miles from Cambridge to Liverpool Street via Bishop's Stortford. A couple of 1938 logs are tabled below.

Cambridge–Liverpool Street, 1938

Miles	Location	8866 5 chs, 157/165 tons 27.1.1938 Driver Bond				8866 6 chs, 165/175 tons 2.20pm Cambridge				Gradients
		Times	Speed			Times	Speed			
0	Cambridge	00.00				00.00				
3.3	Shelford	-				-	sig stand, 1 min 40 secs		L	
6.6	Whittlesford	08.32	66			13.39	63			
10	Great Chesterford	11.39	62			16.49	58/sigs			1/320 R
14	Audley End	15.44	62	¼ E		21.04	sigs		6 L	
15.7	Newport	17.57	pws 36*			23.55	49			
20.1	Elsenham	23.16	40/74			29.43	54/69			1/176 R
22.3	Stansted	25.18	75			32.40	sigs			1/107 F
25.3	Bishops Stortford	28.02	30*	T		37.25			9½ L	
29	Sawbridgeworth	32.42	69			41.36	71			1/340 F, L
38.5	Broxbourne	40.30	79	½ E		49.18	79		7¼ L	1/374, 470 F
45.8	Ponders End	46.14	75			55.04	74/82			1/781 F, L
49.6	Tottenham Hale	49.57	sigs	3 E		58.23	sigs		6¼ L	L
52.7	Hackney Downs	57.21	sigs	¾ E		63.00	pws		5 L	
54.5	Bethnal Green	60.55	41			-				
55.7	Liverpool Street	63.47	(59 net)	1¼ E		68.22	(57 net)		3¼ L	

8783 on the royal train to Wolferton for Sandringham passing Potter's Bar, 8 June 1935. (K. Nunn/LCGB/MLS Collection)

An unrebuilt D15/2, 8862, with Driver Wright, had five coaches, 157/165 tons, in the down direction and completed the non-stop run in 64 minutes 42 seconds (63 net). A rapid climb was made to Bethnal Green in under 3 minutes, 41mph at the summit, 57 before a pws to 32mph at Ponders End, 64 before Bishops Stortford, passed on time, a minimum of 59 on the 1 in 107 climb to Elsenham and 67 on the long 1 in 176 descent after, 57 on the 1 in 130 rise to Audley End and 70 on the falling gradients through Great Chesterfield.

There was little to choose between the performance of the D15s and D16/1 'Super Clauds' as the latter had the same cylinder design, but when the D16/3s came on stream from 1933 onwards their performance matched the best efforts of the 'Super Clauds' when they were new. The D16/3s with piston valves gave a more efficient front-end and these engines with their improved adhesion also gave the B12s a good run for their money. However, as mentioned earlier, these locomotives paid for their more strenuous use with maintenance problems. In the late 1930s they still substituted occasionally if a B12 or B17 was not available for a fast Norwich service. 8808 which had been rebuilt with round-topped boiler and 9½

The other GE section royal engine, 8787, at Ingrave, the summit of Brentwood Bank, with a fast train for Southend, c1932. 8787 and 8783 remained in lined green livery when the other 4-4-0s were painted black.
(F. Moore/MLS Collection)

in piston valves in 1937 gave some distinguished runs in 1938/9, clearing Bethnal Green in 3¼ minutes and Stratford in 8¼ with 375 tons on one occasion, and 8858, a D16/3 slide valve engine, took out 495 tons in July 1938 on the 3.40pm from Liverpool Street, a relief to the main Norwich train. Similar heavy trains were worked on the Cambridge route. The shed allocations of the 37 D15s and 84 D16s at the end of the 1930s just before the war were:

		Change from 1931
Stratford:	24 (12 D16)	-2
Colchester:	8 (5 D16)	-2
Parkeston:	-	-1
Ipswich:	12 (11 D16)	-6
Norwich:	27 (19 D16)	+7
Yarmouth:	8 (6 D16)	+2
Lowestoft:	-	-1
Cambridge:	23 (14 D16)	+4
Bury St Edmunds	2	+2
King's Lynn:	5 (1 D16)	+2
March:	10 (8 D16)	No change
Peterborough E:	2 (2 D16)	-2

D16/3 8876 after rebuilding in March 1935, at Peterborough East with an express for Sheringham and Cromer, c1938. (G. Gillford/R.K. Blencowe/John Scott-Morgan Collections)

The increase in the Norwich allocation reflected more use on the M&GN section taken over in 1936 and on the branches radiating from Norwich and King's Lynn–sWells, Dereham and Hunstanton. The Stratford engines were working heavy commuter trains to both Southend and Bishop's Stortford throughout the 1930s and one train timer, Peter Proud, has placed the logs of many journeys between Liverpool Street and Broxbourne in the Railway Performance Society archives. 8860 and 8866 seemed to spend weeks on the 250-350 ton commuter trains from 1933 to 1936, but maximum speed rarely exceed 55mph in the Down direction and upper 50s in the Up. Only one run was logged at 60mph – 8900 itself with 63mph on a 225 ton train on the morning up run in February 1937. Timekeeping was reliable if unchecked. The schedule for the heavy 5.10pm or 6.30pm evening home runs was 27 minutes for the 17 miles including Bethnal Green Bank and restricted speed between Hackney Downs and Tottenham and then the gradual grades against the engine. The fastest down run noted was with 8866 after rebuilding as a D16/3 with piston valves in March 1933, which completed the run on the 6.30pm with 325 tons in 25 minutes 50 seconds in October 1933. This engine like many other piston valve engines, was an early withdrawal (in 1945) with cracked frames. Peter Proud noted the following D15 and D16s on these services between 1931 and 1939: 8797, 8804, 8816, 8821, 8823, 8824, 8830, 8835, 8836, 8848,

8849, 8857, 8860, 8865, 8866, 8882, 8890, 8893, 8896, 8899, 8900.

In the war years, the D15 and D16s, painted plain black, became common user engines, including the two former royal engines, though 8868 was kept in good shape for any royal duties. Southend was an evacuation area and Stratford 'Clauds' were used to run specials there. 8787 was put on a 17-coach 4.15pm King's Cross–Leeds train at Cambridge after a freight derailment at Arlesey in 1943 caused a diversion and took an hour to cover the 29 miles from Peterborough to Grantham, despite having plenty of steam – the climb to Stoke Tunnel clearly took its toll. When 1671 was rebuilt as a B2 in 1946 and became the royal engine for the Sandringham trains, 8783 transferred to King's Lynn and although 8787 stayed at Cambridge it was not retained for special working.

The first 'Claud' to be withdrawn was 8866 and a further three were condemned before the remaining 117 were transferred to BR stock, three of which were piston valve engines including the pioneer 2500 (8900) *Claud Hamilton* whose name was transferred to 2546 ex-works from Stratford in August 1947. The LNER Eastern Section had received forty-five new Thompson B1s releasing B12s and B17s for secondary work and a steady withdrawal of the remaining D15s and the D16s took place from 1950 onwards. They were very common on the M&GN section and the branches from King's Lynn. In the London area a few were still used on the King's Cross buffet expresses in 1948/9 and according to Peter Proud's records, Stratford D16s (and a D15, numbered 7764 between 1943 and 1946, later 2505) continued to head his commuter runs to and from Broxbourne in the immediate post-war years. The allocation at the start of nationalisation in January 1948 was:

		Change from 1939
Stratford:	5 (3 D16s)	-19
Colchester:	7 (all D16s)	-1
Ipswich:	7 (all D16s)	-5
Norwich:	22 (all D16s)	-5
Yarmouth:	15 (all D16s)	+7
Lowestoft:	2 (both D16s)	+2
Cambridge:	16 (all D16s)	-7
Bury St Edmunds:	4 (2 D16s)	+2
King's Lynn:	14 (7 D16s)	+9
March:	9 (all D16s)	-1
South Lynn:	6 (5 D16s)	+6
Melton Constable:	6 (4 D16s)	+6
Yarmouth Beach:	4 (all D16s)	+4

The Oxford–Cambridge line was placed fully within the Eastern Region on nationalisation and D16s and B12s worked through to Oxford and after the LMR had tried to use ex LMS 4-4-0s unsuccessfully on the CLC lines, 62535 joined the former GC D9s in 1949 and was followed by seven more in early 1950, based at Trafford Park.

They were stored by late 1951 and returned to the Eastern Region at the end of the summer of 1952. Apparently, they were well liked by local crews but there were maintenance problems obtaining spare parts from Stratford Works. By the end of 1954 the class was down to ninety, all D16/3s. The allocation in January 1955 was:

		Change from 1948
Stratford:	-	-5
Colchester:	-	-7
Ipswich:	2	-5
Norwich:	17	-5
Yarmouth:	11	-4
Lowestoft:	-	-2
Cambridge:	17	+1
Bury St Edmunds:	6	+2
King's Lynn:	14	No change
March:	9	No change
Spital Bridge (Peterborough M&GN):	7	+7
South Lynn:	-	-6
Melton Constable:	7	+1
Yarmouth Beach:	-	-4

Although none was allocated to Lincoln, they had appeared there for many years working from March, but in March 1957 five were allocated there for services to Newark, Nottingham and Derby, replacing the remaining 'Director' D11s. They were withdrawn, however, by the end of 1958. Only seventeen D16s were still in service at the beginning of 1959, allocated as follows:

Norwich:	62511, 62517 (to March in June), 62524, 62540, 62544, 62597 (to Spital Bridge in August), 62612 (to Spital Bridge in April)
March:	62529, 62589, 62618

D16/2 2553, a Norwich engine. c1947. This engine was not rebuilt to class D16/3 until September 1949.
(MLS Collection)

D16/3 62535, one of the locomotives transferred to the Cheshire Lines Committee after the Second World War, with a Liverpool Central–Manchester Central stopping train near Risley Moss, 19 March 1950.
(J.D. Darby/MLS Collection)

King's Lynn: 62606
Yarmouth: 62570 (to March in June), 62604 (to Lowestoft in September), 62613 (to Spital Bridge in April)

The last D16s were 62524 withdrawn from Norwich in March 1960, 62604 from Lowestoft in February 1960, and the last of all, 62613 from March in November 1960. It starred on the Down 'Fenman' between Cambridge and King's Lynn on Easter Monday 1960. Its last known working was the 4.40pm Norwich–Peterborough from March on 15 September. The oldest was 62524 which had a career of exactly 58 years.

62536 at Southport with a return excursion via the CLC lines, 2 June 1951.
(MLS Collection)

D15 62507 (former S46 1896 built in 1900), superheated in 1915 and fitted with extended smokebox as a D15/2 in 1930, at Hunstanton with a train for Liverpool Street, August 1951. It is still fitted with a watercart tender and was withdrawn in April 1952.
(MLS Collection)

D16/3 62612 at Cambridge with a northbound local train for Ely, 1950. The revised shape of the front end of the frame is very conspicuous here. 62620 is also one of the D16s that retained its decorative valances. (MLS Collection)

D16/3 62620 with a goods train at Melton Constable, 25 August 1952. (J.D. Darby/ MLS Collection)

D16/3 62578 leaving Melton Constable with a train for Cromer, 9 May 1951.
(N. Fields/MLS Collection)

D16/3 62577 was rebuilt as a D16/3 as late as 1949. It is seen here at Dereham, 7 June 1952.
(J.A. Peden.MLS Collection)

D16/3 62592 pilots E4 2-4-0 62787 on a local at Dereham, 1955. (N. Harrop/MLS Collection)

D16/3 62562 at Wisbech with the 3.30pm Sundays Only train to March, 4 October 1953. (MLS Collection)

D16/3 62561 at Weybourne (now the North Norfolk Heritage Railway) with a Melton Constable–Cromer train, 9 May 1956.
(T.K. Widd/MLS Collection)

D16/3 62571 at Sheffield Victoria with a stopping train for Lincoln, 19 August 1957.
(P.J. Hughes/MLS Collection)

Preservation

The Claud Hamilton Locomotive Group regretted that no Great Eastern 4-4-0 of class D15 or D16 was preserved and formed a charity in 2016 to raise funds to construct a new D16/2, a replica of the LNER royal engine of the 1920s and 1930s, 8783. Its homebase is intended to be the Whitwell and Reepham Railway. As yet, the construction has not started and no date for its completion has been set. It is intended to name the new build engine *Phoenix*.

Model

Hornby produced a model of the D16/3 in 2017 (catalogue R 3303) and I purchased a BR plain black version, lightly weathered as I remember seeing them at Oxford in the mid-1950s arriving from Cambridge. (David Maidment)

One of the H88/D16s built in 1923 and rebuilt as a class D16/2 in June 1928 at King's Cross, c1930. At this time 8783 was a regular 'royal' engine for the royal family's visits to Sandringham. (MLS Collection)

Chapter 5
THE MIDLAND & GREAT NORTHERN RAILWAY 4-4-0s

The Midland and Great Northern Railways assumed the working of trains on the Eastern & Midlands Railway west of King's Lynn from 1889 and the complete system of Peterborough to Melton Constable, Cromer and Yarmouth from 1893, forming the M&GN Joint Committee to run the railway. The LNER took over sixty of the eighty-five M&GN locomotives on 1 October 1936 and promptly withdrew a further twenty, leaving forty to be taken into LNER stock and receive LNER class classification. Melton Constable Works was closed at the end of 1936 and major overhauls transferred to Stratford.

M&GN 'A' class rebuilds

Fifteen outside cylinder 4-4-0s were built by Beyer Peacock between 1882 and 1886, numbered 21–35 and were identified as class 'A'. They were rebuilt with standard Midland Railway boilers between 1895 and 1909 and the eight numbered 21–28 were rebuilt a second time between 1914 and 1927, and known as class 'A Rebuild'. Extended smokeboxes were fitted to Nos. 31-35 when they were reboilered between 1907 and 1909 and the rest of the class were treated similarly subsequently. Five of the 'A Rebuilds', 23 and 25–28, were taken into LNER stock in 1936, the remainder having been withdrawn earlier. The dimensions of the five that became owned by the LNER were:

Cylinders
  (2 outside):   17 x 24in

Stephenson motion with slide valves
Coupled wheel
  diameter: 6ft 0in
Bogie wheel
  diameter: 3ft 0in
Boiler pressure: 160lbs psi
Heating surface: 1,072.38sq ft
Grate area: 16sq ft
Axleload: 15 tons 12 cwt
Weight (Engine):   41 tons 3 cwt
        (Tender):   36 tons 9 cwt
        (Total):   77 tons 12 cwt
Water capacity: 3,000 gallons
Coal capacity: 3 tons
Tractive effort: 13,101lbs

Nos. 23, 26 and 27 were withdrawn almost immediately in early 1937, 28 the following

M&GN Beyer Peacock class 'A' Rebuild, No.24 built in 1882, rebuilt with a Midland type boiler, seen here c1935, and renumbered 25 when rebuilt with a Midland boiler the second time in 1936 (an amalgam of 24 and 25 on 24's frame). It was the last survivor of the class, withdrawn in May 1941. (MLS Collection)

M& GN class 'C' No.28, later LNER D52, at Melton Constable, c1930. It was taken into LNER stock in 1936 and withdrawn in February 1938. (R.K. Blenkowe/John Scott-Morgan Collections)

year and 25, the last survivor, in May 1941. 24 had been withdrawn at the time of the LNER take-over and its frame coupled with the boiler of 25, the overhauled engine taking this number. It was renumbered 025 during a Stratford Works visit in June 1937 and was then repainted in LNER unlined black livery.

The locomotives worked between Lynn, Norwich and Yarmouth. After the arrival of the Midland 'C' class in 1894, they were displaced and worked local services all over the M&GN system. Latterly they were used on local passenger and freight services in the Spalding/Bourne area.

M&GN Beyer Peacock class 'A' Rebuild No.26, built in 1883, and rebuilt with Midland type 'C' boilers in 1904 and 1923, at Spalding with a local train, c1930. 26 was withdrawn by the LNER just one month after taking it into stock in November 1936. (MLS Collection)

LNER D52

In 1894, the Committee procured from the Sharp Stewart Company some Johnson inside cylinder 4-4-0s of its own to dispense with the use of Midland and Great Northern locomotives – engines that were similar to the '1808' class being built at the time for both the Midland and Somerset & Dorset Railways. Twenty-six of these class 'C' locomotives were built in 1894, numbered 1–7, 11–14, 17, 18, 36–39 and 42–50. Seven more, 51–57, were delivered in 1896 and another seven, 74–80, the latter built by the Beyer, Peacock Company, in 1899. Their dimensions were:

Cylinders (2 inside):	18 ½ x 26in
Coupled wheel diameter:	6ft 6 ½ in
Bogie wheel diameter:	3ft 3 ½ in
Boiler pressure:	160lbs psi
Heating surface:	1,078sq ft
Grate area:	17.5sq ft
Axleload:	16 tons
Weight (Engine):	42 ton 18 cwt
(Tender):	33 tons 11 cwt
(Total):	76 tons 9 cwt
Water capacity	2,950 tons
Coal capacity	3 tons
Tractive effort	15,416lbs

The only differences to the Midland '1808s' were the half-inch larger diameter cylinders and deeper frames. The Beyer, Peacock engines to the same design were the blueprint for the Midland Railway's '2581' class built by the same company in the same year.

Twenty-three of the locomotives remained unaltered apart from

Midland & Great Northern class 'C' 39, built in May 1894 seen here c1905. It was later rebuilt with G7 Belpaire boiler in 1908 and classified by the LNER as a D54. It was withdrawn in February 1937. (MLS Collection)

A M&GN 4-4-0 class 'C', No.14, designed by S.W. Johnson at Derby similar to the Midland '1808' class and the Somerset & Dorset class 'A', constructed by Sharp, Stewart & Co. in 1894, taken at Cromer, c1905. 14 remained unrebuilt and was withdrawn by the LNER as a D52 in February 1937. (Locomotive Publishing Co./MLS Collection)

boiler replacements of similar dimensions although they received Midland type smokebox doors and detailed alterations to cab and chimneys. They were taken into LNER stock in October 1936 and classified D52. The immediate LNER allocation was:

Melton Constable:	4, 5, 14, 17, 37, 38, 43, 47, 48, 80
Yarmouth Beach:	1, 3, 12
South Lynn:	7, 18, 74, 75, 78
New England:	13, 42, 76
Peterborough East:	11, 79

The last D52 was No.38, withdrawn from Melton Constable in September 1943.

Johnson 4-4-0, No.14, built in 1894 as above, but reboilered with small secondhand Derby boiler with extended smokebox, Midland style smokebox door and tall narrow chimney in 1934, withdrawn in February 1937. The LNER classified these 1894 locomotives as D52. (Photomatic/ MLS Collection)

M&GN class 'C' No.18 arrives at Sutton bridge with a summer Saturday Yarmouth holiday express, passing 'A' Rebuild No.28 of 1883 on a local passenger train, c1930. (R.K. Blenkowe/ John Scott-Morgan Collections)

Johnson 4-4-0 M&GN No.1, built in 1894 and rebuilt with a Midland/LMS small boiler in 1932, on a stopping passenger service on the M&GN system, c1935. It was withdrawn in November 1937. (Locomotive & General/MLS Collection)

M&GN/LNER D52 047 at Yarmouth Beach 4 September 1938. It was withdrawn in June 1942. (MLS Collection)

LNER D53 & D54

Seven of the Johnson 'C' engines – 2, 6, 36, 44, 49, 60 and 77 – were rebuilt with Midland G6 boilers between 1929 and 1931 and were classified by the LNER as D53s. As early as 1908, two (39 and 55) were reboilered with the 'H' boiler and between 1910 and 1925 these two and eight more – 45, 46 and 51-57 – were rebuilt with Belpaire G7 boilers and in 1936 were classified as D54s. The D53s weighed 30 cwt heavier, the D54s had boiler pressure of 175lbs psi, heating surface of 1,384sq ft and a larger grate area of 21sq ft and weighed 83 tons 9 cwt. Tractive effort was thereby increased to 16,862lbs. The D54s were mainly on the Peterborough–Cromer and Leicester–Yarmouth and Lowestoft through services, the D53s working the stopping passenger services, fish and goods trains. At the LNER take-over, the allocation of the D53s and D54s was:

Melton Constable:	2 (D53), 39, 54 (D54)
Yarmouth Beach:	6 (D53), 46, 56 (D54)
South Lynn:	45, 51-53, 55, 57 (D54)
New England:	49, 77 (D53)
Peterborough East:	36, 44, 50 (D53)

The LNER reviewed the stock they had inherited and withdrew thirteen D52s, one D53 and three D54s between November 1936 and November 1937. They renumbered all the M&GN engines on the 'duplicate' list with the '0' prefix. The last D54s were 055 and 056 withdrawn in November 1943 and the last of all, D53s 050 and 077, withdrawn in January 1945 before they received their allocated LNER numbers of 2052 and 2054. (06 allocated 2053 was withdrawn in 1944.)

I cannot trace any runs by the former M&GN engines after their absorption into LNER stock. The only logs I can find in more recent M&GN times were two runs with D53 No.77 in the mid-1930s. Both were from Melton Mowbray to Nottingham, 18.2 miles, with no more than three coaches weighing 85 tons. On 22 August 1935, 77 completed the run in 27 minutes 5 seconds which included a signal check to 5mph almost immediately after leaving Melton and a p-way slack between Edwalton and Nottingham. Top speed at Widmerpool was 63mph. On 4 February 1937 the same engine and load were much brisker and took just 22½ minutes unchecked, with 60mph at Old Dalby, 66 at Widmerpool, 57 at Plumtree and 62 at Edwalton. The schedule was 24 minutes.

M&GN class 'C' No.77 as built in November 1899 and before rebuilding as a D53 in 1930, seen here c1912. (F. Moore/MLS Collection)

The Midland & Great Northern Railway 4-4-0s • 213

No.77 after rebuilding in 1930 with a G6 boiler as LNER class D53, at Nottingham on 13 June 1935. 77 was one of the last M&GN engines in active use on the LNER and was withdrawn in January 1945.
(G.A. Barlow/MLS Collection)

M&GN class 'C' No. 45 built in June 1894 and rebuilt with Midland G7 Belpaire boiler in 1909, seen here shortly afterwards. It was classified a D54 by the LNER and withdrawn in November 1936.
(Locomotive Publishing Co./MLS Collection)

M&GN class 'C' No.56 built in 1896, rebuilt with G7 Belpaire boiler in 1912 and withdrawn by the LNER as a D54 in November 1943. (MLS Collection)

Johnson 1896 4-4-0 M&GN No.54, rebuilt in 1914 with Deeley G7 boiler, extended smokebox and Belpaire firebox, c1930. It was withdrawn as LNER D54 class in October 1939. (Locomotive & General/ MLS Collection)

Johnson 4-4-0 No.77 rebuilt with Belpaire G6 boiler in December 1930, a D53 renumbered 077 on the duplicate list when the LNER took responsibility for M&GN locomotives in October 1936, seen here approaching Nottingham a few months earlier, 19.5.1936. It was withdrawn in January 1945. (G.A. Barlow/MLS Collection)

M&GN No.53, rebuilt in 1910 with Deeley G7 boiler, at Thurmaston with a Birmingham–Yarmouth train, 4 June 1911. It was classified as a D54 by the LNER and withdrawn in January 1940. (MLS Collection)

LNER D53 No.06, rebuilt from an M&GN class 'C' in 1930, on a local freight train, c1938. It was withdrawn in March 1944.
(MLS Collection)

D54 052 with Deeley G7 boiler, after absorption by the LNER in 1936, at Yarmouth Beach shed, 4 September 1938.
(MLS Collection)

COLOUR SECTION

Left: **The rebuilt** Ivatt D3 2000, formerly 4075, rebuilt and repainted for general management saloon and other special duties, at Grantham, c1947. (David Williams colourisation)

Below left: **D10 GC Director** 62656 *Sir Clement Royds* at Manchester Central with an unidentified B1, 1952. (MLS Collection)

Below right: **D11/1** 62667 *Somme* on an RCTSC rail tour at Wintersett Junction, 7 June 1953. (Colour Rail)

D11/1 62667 *Somme* with a Nottingham–Derby train at Trent Junction, July 1955. (Colour Rail)

Below left:
D11/1 62668 *Jutland* at Nottingham Victoria with a GW 'Siphon' bogie van, May 1958, May 1958. (Colour Rail)

Below right:
D11s, 62660 *Butler-Henderson* and **62665** *Mons* on Sheffield Darnall shed, September 1958. (K.R. Pirt/MLS Collection)

D11 62669 *Ypres* departing from Sheffield Victoria with a stopping train to Doncaster, a K3 2-6-0 in the background, c1958. (MLS Collection)

Below left: **D11 62660** *Butler-Henderson* with an RCTS railtour photostop at Hare Park, 21 September 1958. (Transport Online Collection)

Below right: **D11 62660** *Butler-Henderson* with an RCTS railtour at Calder Bridge Junction, 21 September 1958. (Transport Online Collection)

62669 *Ypres* at Nottingham Victoria with a local train for Sheffield, a B1 in the background, 15 August 1960. (Colour Rail)

Below left: D11/2 62677 *Edie Ochiltree* in LNER lined green as applied to a handful of Scottish D11/2s after the 1946 renumbering, at Haymarket, 1949. (Colour Rail)

Below right: A line of D11/2s in store at Longniddry, led by 62673 *Evan Dhu*, February 1958. (Colour Rail)

Above left: **A Great** Eastern D13 in GER blue livery crossing the Trowse swing bridge near Norwich, c1912. (Ian MacCabe colourisation)

Above right: A Great Eastern D13 picking the single line tablet up at Heacham, c1930. (Ian MacCabe colourisation)

Left: **One of** the LNER's two D16/2 locomotives used for royal train journeys to Sandringham, 8783, c1938. (Ian MacCabe colourisation)

Below: **The other** 'royal' D16/2, 8787, sweeps into Welwyn North station alongside the station's horse shunter, with a Cambridge–King's Cross buffet express, c1938. (Ian MacCabe colourisation)

D15/2 8893 at Bishops Stortford, 1938. It was built as 1893 in May 1900 and withdrawn still as a D15 numbered 2504 in 1948. (Colour Rail)

Below left: **D16/3 2573** waits to depart from an unidentified station, c1947. (Ian MacCabe colourisation)

Below right: **D16/3 62574** at Sudbury, February 1954. (Colour Rail)

Colour Section • 223

Above left:
D16/3 62530 at Cambridge, May 1959.
(K.R. Pirt/MLS Collection)

Above right:
D16/3 62571 waits at xxxx station with a train for xxxx, c1956.
(Transport Online Collection)

D16/3 62596 with the 9am Gorleston–York at Gunton, August 1957.
(Colour Rail)

Above left: D16/3 62618, formerly 8787, at its home depot, Cambridge, c1957.
(Ian MacCabe colourisation)

Above right: D16/3 62517 at Caister on the 9.30an SO Yarmouth Beach–Derby, July 1957.
(Colour Rail)

D16/3 62614, formerly 8783, one of two D16s painted in LNER apple green livery post the Second World War and maintaining the livery in the early days of nationalisation, seen here on shed, c1951.
(Ian MacCabe colourisation)

APPENDIX

The D1 4-4-0
Dimensions – see page 11 & 12

Statistics

No.	Built	Renumbered LNER		1946 scheme		BR	Withdrawn
51	3/1911	3051	2/25	(2202)		-	2/1946
52	4/1911	3052	9/24	2203	1/1947	62203 3/1948	8/1950
53	4/1911	3053	10/25	(2204)		-	3/1946
54	4/1911	3054	8/1924	2205	6/1946	(62205)	11/1948
55	5/1911	3055	2/1925	(2206)		-	6/1946
56	5/1911	3056	8/1924	2207	1/1947	(62207)	11/1948
57	5/1911	3057	2/1925	2208	12/1946	62208 4/1948	7/1950
58	5/1911	3058	7/1925	2209	6/1946	(62209)	11/1950
59	6/1911	3059	2/1924	2210	1/1947	-	12/1947
60	6/1911	3060	2/1925	(2211)		-	12/1946
61	6/1911	3061	7/1924	2212	4/1946	-	7/1947
62	6/1911	3062	9/1925	(2213)		-	6/1946
63	6/1911	3063	7/1925	2214	6/1946	(62214)	10/1949
64	7/1911	3064	6/1925	2215	6/1925	62215 3/1948	2/1950
65	8/1911	3065	1/1925	2216	11/1946	-	9/1947

The D2 4-4-0 (GNR D1)
Dimensions – see page 19

Statistics

No.	Built	Renumbered LNER		Superheated	1946 scheme	BR	Withdrawn
1321	6/1898	4321	4/1925	-	2152	10/1946 (62152)	1/1949
1322	6/1898	4322	12/1924	-	-	-	7/1939
1323	6/1898	4323	8/1925	-	2153	9/1946 (62153)	4/1949
1324	6/1898	4324	8/1924	-	2154	12/1946 (62154)	11/1950

No.	Built	Renumbered LNER	Superheated	1946 scheme	BR	Withdrawn
1325	7/1898	4325	10/1925	-	-	-	12/1937
1326	10/1898	4326	1/1926	-	2155	9/1946 (62155)	2/1948
1327	10/1898	4327	1/1925	2/1935	2156	11/1946 (62156)	1/1949
1328	10/1898	4328	11/1924	6/1935	-	-	10/1937
1329	11/1898	4329	11/1924	-	2157	11/1946 (62157)	4/1948
1330	11/1898	4330	2/1924	5/1935	(2158)	-	3/1946
1331	11/1898	4331	1/1925	-	2159	10/1946 -	7/1947
1332	11/1898	4332	6/1924	3/1930	2160	2/1946 (62160)	10/1948
1333	11/1898	4333	3/1924	-	2161	4/1946 (62161)	7/1950
1334	11/1898	4334	2/1925	5/1928	-	-	9/1936
1335	11/1898	4335	10/1925	-	2162	9/1946 -	7/1947
1336	9/1899	4336	7/1924	-	-	-	5/1937
1337	10/1899	4337	1/1925	3/1935	2163	7/1946 (62163)	10/1948
1338	10/1899	4338	4/1925	2/1935	(2164)	-	2/1946
1339	10/1899	4339	10/1925	11/1929	2165	11/1946 (62165)	3/1949
1340	10/1899	4340	4/1925	-	2166	11/1946 -	7/1947
1361	11/1899	4361	9/1925	11/1929	2167	12/1946 (62167)	2/1949
1362	10/1899	4362	10/1924	-	-	-	6/1937
1363	10/1899	4363	12/1924	5/1936	-	-	6/1939
1364	11/1899	4364	12/1924	-	2168	-	12/1947
1365	11/1899	4365	11/1924	3/1935	2169	9/1946 (62169)	7/1948
1366	10/1900	4366	1/1925	-	2170	3/1946 -	8/1947
1367	10/1900	4367	3/1926	10/1935	-	-	6/1939
1368	10/1900	4368	7/1924	-	2171	7/1946 -	4/1947
1369	10/1925	4369	10/1925	-	2172	7/1946 62172 3/48	6/1951
1370	11/1900	4370	3/1925	-	2173	10/1946 (62173)	5/1950
1371	11/1900	4371	10/1924	-	2174	8/1946 -	9/1947
1372	11/1900	4372	2/1924	-	-	-	12/1938
1373	11/1900	4373	10/1924	-	2175	8/1946 (62175)	11/1948
1374	12/1900	4374	6/1924	-	2176	7/1946 -	1/1947
1375	12/1900	4375	7/1924	-	-	-	12/1939
1376	12/1900	4376	1/1925	-	-	-	11/1937
1377	1/1901	4377	7/1924	1/1935	2177	8/1946 (62177)	10/1949
1378	1/1901	4378	2/1925	2/1928	-	-	6/1937

No.	Built	Renumbered LNER	Superheated	1946 scheme	BR	Withdrawn
1379	1/1901	4379 6/1926	2/1935	2178	9/1946 -	7/1947
1380	11/1900	4380 4/1925	-	2179	9/1946 (62179)	3/1949
1381	1/1901	4381 3/1925	4/1914	2180	7/1946 (62180)	5/1950
1382	11/1900	4382 4/1925	3/1930	-	-	5/1937
1383	12/1900	4383 11/1925	-	2181	7/1946 (62181)	11/1950
1384	12/1900	4384 2/1925	-	(2182)	-	3/1946
1385	12/1900	4385 6/1924	-	2183	7/1946 -	7/1947
1386	1/1903	4386 4/1925	-	-	-	3/1939
1387	2/1903	4387 2/1924	10/1935	2184	9/1946 -	12/1947
1388	2/1903	4388 7/1924	-	2185	7/1946 -	7/1947
1389	2/1903	4389 9/1924	-	-	-	4/1937
1390	5/1903	4390 2/1925	-	2186	8/1946 -	5/1947
1391	6/1903	4391 3/1925	5/1935	-	-	8/1938
1392	4/1903	4392 1/1925	11/1929	2187	10/1946 (62187)	10/1948
1393	2/1903	4393 6/1924	-	2188	10/1946 (62188)	10/1949
1394	5/1903	4394 1/1925	3/1936	2189	8/1946 (62189)	11/1948
1395	3/1903	4395 9/1926	-	2190	7/1946 (62190)	9/1949
1396	8/1907	4396 4/1924	-	-	-	8/1938
1397	9/1907	4397 9/1925	-	-	-	11/1938
1398	10/1907	4398 4/1924	-	2191	9/1946 -	6/1947
1399	10/1907	4399 6/1925	5/1928	(2192)	-	5/1946
1180	10/1907	4180 11/1924	-	2193	12/1946 (62193)	6/1949
41	4/1909	3041 7/1924	-	2194	4/1946 (62194)	6/1949
42	4/1909	3042 3/1924	-	2195	12/1946 (62195)	2/1948
43	4/1909	3043 7/1924	-	-	-	6/1938
44	5/1909	3044 12/1924	-	2196	1/1947 -	5/1947
45	5/1909	3045 7/1925	-	2197	12/1946 (62197)	1/1949
46	6/1909	3046 8/1924	-	-	-	11/1938
47	6/1909	3047 8/1925	-	2198	12/1946 (62198)	8/1948
48	6/1909	3048 3/1925	10/1935	2199	4/1946 (62199)	7/1949
49	6/1909	3049 9/1924	-	2200	7/1946 -	7/1946
50	6/1909	3050 1/1925	7/1937	2201	12/1946 -	7/1947

*Rebuilt from D3

No.	Built	Renumbered LNER	Superheated	1946 scheme	BR	Withdrawn
1305	10/1923*	4305 12/1925	-	2150	11/1946 (62150)	5/1949
1320	6/1926*	4320 4/1925	-	2151	10/1946 (62151)	4/1949

The D3 & D4 4-4-0 (GNR D2)
Dimensions – see page 30

Statistics

No.	Built	Rebuilt to D3	Renumbered LNER	1946 scheme		BR	Withdrawn
400	12/1896	9/1920	3400 5/1925	2115	8/1946	-	9/1947
1071	5/1897	12/1917	4071 1/1925	2116	8/1946	(62116)	10/1948
1072	5/1897	4/1913	4072 1/1925	-	-	-	12/1935
1073	6/1897	4/1916	4073 7/1924	(2117)	-	-	2/1946
1074	6/1897	2/1917	4074 10/1924	2118	8/1946	-	11/1946
1075	6/1897	8/1916	4075 11/1924	2000	10/1944	62000 1/1950	10/1951
1076	6/1897	1/1919	4076 10/1924	-	-	-	6/1936
1077	6/1897 D4	6/1923	4077 3/1925	-	-	-	10/1937
1078	6/1897	10/1917	4078 5/1925	-	-	-	12/1935
1079	6/1897 D4	1/1926	4079 3/1925	-	-	-	10/1937
1080	6/1897	3/1918	4080 8/1925	2120	9/1946	-	8/1947
1301	10/1897	7/1916	4301 1/1925	2121	9/1946	-	8/1947
1302	10/1897	3/1916	4302 1/1925	2122	9/1946	(62122)	2/1948
1303	11/1897	4/1921	4303 11/1925	2123	7/1946	(62123)	12/1949
1304	11/1897	2/1917	4304 7/1926	-	-	-	11/1942
1305	11/1897	4/1913	4305 12/1925	2150	11/1946	(62150) D2	10/1923
1306	11/1897	3/1913	4306 5/1924	2124	9/1946	(62124)	11/1948
1307	11/1897	6/1917	4307 5/1925	2125	5/1946	(62125)	8/1949
1308	12/1897	6/1916	4308 8/1925	-	-	-	10/1935
1309	12/1897	2/1918	4309 3/1926	2126	9/1946	(62126)	8/1948
1310	12/1897	3/1917	4310 4/1925	2127	8/1946	-	5/1947
1311	3/1898	12/1918	4311 7/1924	2128	11/1946	(62128)	12/1949
1312	3/1898	4/1917	4312 7/1924	(2129)	-	-	2/1946
1313	4/1898 D4	5/1924	4313 5/1924	-	-	-	12/1935
1314	5/1898	1/1918	4314 1/1925	-	-	-	6/1937
1315	4/1898	5/1915	4315 1/1926	2130	9/1946	-	4/1947
1316	4/1898	2/1914	4316 2/1925	2131	12/1946	62131 3/1948	10/1949
1317	5/1898	2/1913	4317 6/1925	2132	9/1946	(62132)	12/1950
1318	5/1898	9/1920	4318 6/1925	2133	10/1946	(62133)	8/1949
1319	6/1898	10/1916	4319 2/1924	2134	9/1946	-	12/1946

No.	Built	Rebuilt to D3	Renumbered LNER	1946 scheme		BR	Withdrawn
1320	6/1898	10/1920	4320 4/1925	2151	10/1946	(62151) D2	6/1926
1341	12/1898	8/1915	4341 2/1924	-	-	-	7/1937
1342	11/1898	11/1917	4342 9/1925	-	-	-	11/1936
1343	12/1898	4/1917	4343 1/1925	2135	6/1946	62135 6/1948	2/1950
1344	12/1898	12/1917	4344 9/1924	2136	9/1946	-	11/1947
1345	12/1898	2/1917	4345 2/1924	2137	8/1946	(62137)	1/1949
1346	12/1898	12/1917	4346 1/1925	2138	9/1946	-	8/1947
1347	12/1898	7/1914	4347 10/1924	2139	11/1946	(62139)	6/1949
1348	12/1898	5/1915	4348 4/1925	2140	9/1946	(62140)	6/1950
1349	12/1898	10/1915	4349 8/1925	2141	10/1946	-	11/1946
1350	12/1898	5/1916	4350 1/1925	2142	10/1946	-	4/1947
1351	11/1899	9/1918	4351 10/1924	2143	10/1946	(62143)	3/1948
1352	11/1899	7/1914	4352 2/1926	2144	10/1946	(62144)	8/1948
1353	11/1899	4/1917	4353 5/1924	-	-	-	4/1936
1354	12/1899	3/1913	4354	-	-	-	10/1937
1355	12/1899	8/1916	4355 3/1924	2145	11/1946	(62145)	1/1949
1356	12/1899 D4	9/1927	4356 11/1924	(2146)	-	-	6/1946
1357	12/1899	9/1916	4357 10/1925	2147	2/1946	-	6/1947
1358	12/1899 D4	6/1928	4358 7/1925	-	-	-	8/1937
1359	12/1899	11/1912	4359 1/1925	2148	7/1946	(62148)	11/1950
1360	12/1899 D4	9/1927	4360 11/1924	-	-	-	10/1935

GC class '11' (D5)
Dimensions – see pages 43 & 44

Statistics

No.	Built	Reboilered	Renumbered	Superheated	Withdrawn
694	7/1895	1913	5694 11/1925	11/1925	10/1932
695	8/1895	1911	5695 8/1926	8/1926	7/1931
696	10/1895	1906	5696 6/1924	-	7/1930
697	10/1895	1906	5697 12/1924	-	10/1932
698	11/1895	1905	5698 12/1926	-	10/1931
699	12/1895	1911	5699 11/1925	-	3/1933

GC class '11A' (D6)
Dimensions – see page 47

Statistics

No.	Built	Built by	LNER No.	1946 No.	Westinghouse	Superheated	Withdrawn
268	9/1897	Gorton	5268	-	-	1/1915	7/1933
269	12/1897	Gorton	5269	-	-	5/1927	11/1932
270	12/1898	Gorton	5270	(2103)	-	1/1915	4/1945
852	4/1898	Gorton	5852	-	-	10/1921	6/1938
853	5/1898	Gorton	5853	(2100)	-	4/1914	4/1946
854	6/1898	Gorton	5854	-	-	5/1912	5/1931
855	7/1898	Gorton	5855	2101	-	6/1914	12/1947
856	8/1898	Gorton	5856	-	-	6/1921	1/1939
857	8/1898	Gorton	5857	-	1902-1931	1/1916	8/1931
858	9/1898	Gorton	5858	-	-	8/1913	11/1931
859	10/1898	Gorton	5859	(2102)	1903-1932	10/1916	9/1945
860	11/1898	Gorton	5860	-	-	7/1913	11/1931
861	11/1898	Gorton	5861	-	-	2/1916	4/1931
862	12/1898	B. Peacock	5862	-	-	5/1912	5/1933
863	12/1898	B. Peacock	5863	-	-	12/1919	4/1937
864	1/1899	B. Peacock	5864	-	-	6/1912	10/1938
865	1/1899	B. Peacock	5865	2104	-	5/1912	12/1946
866	1/1899	B. Peacock	5866	-	-	4/1927	9/1930
867	1/1899	B. Peacock	5867	-	-	7/1929	9/1932
868	1/1899	B. Peacock	5868	-	-	6/1913	4/1933
869	2/1899	B. Peacock	5869	-	1903-1934	4/1914	6/1943
870	2/1899	B. Peacock	5870	-	-	7/1912	6/1931
871	2/1899	B. Peacock	5871	(2105)	-	11/13-12/23	6/1946
872	2/1899	B. Peacock	5872	-	-	2/1913	3/1933
873	3/1899	B. Peacock	5873	-	-	2/1913	6/1930
874	3/1899	B. Peacock	5874	2106	-	7/1914	12/1947
875	3/1899	B. Peacock	5875	-	-	2/1913	11/1935
876	3/1899	B. Peacock	5876	-	1903-1932	3/1913	3/1939
877	3/1899	B. Peacock	5877	-	-	4/1914	7/1932
878	3/1899	B. Peacock	5878	-	-	3/1934	10/1938
879	4/1899	B. Peacock	5879	(2107)	-	5/1919	11/1943
880	4/1899	B. Peacock	5880	-	-	6/1923	4/1939
881	4/1899	B. Peacock	5881	-	-	12/1912	12/1932

GC class '2 & 2A' (D7)
Dimensions – see page 56

Statistics

No.	Built	Builder	Reboilered	Renumbered		Withdrawn	
561	11/1887	Kitson & Co	12/1916	5561	10/1926	9/1928	
562	4/1890	Gorton	12/1912	5562	7/1924	8/1935	
563	4/1890	Gorton	5/1910	5563	9/1924	10/1929	
564	7/1890	Gorton	7/1910	5564	1/1924	10/1926	
565	8/1890	Gorton	11/1911	5565	10/1924	4/1933	
566	9/1890	Gorton	2/1916	5566	12/1925	8/1931	
567	12/1890	Gorton	6/1918	5567	7/1925	9/1931	
682	11/1891	Gorton	6/1909	5682	11/1925	12/1931	
683	12/1891	Gorton	10/1917	5683	5/1925	4/1937	
684	11/1891	Gorton	5/1912	5684	3/1926	6/1939	
685	1/1892	Gorton	1/1910	5685	6/1924	7/1930	
686	3/1892	Gorton	4/1910	5686	8/1924	8/1933	
687	5/1892	Gorton	6/1909	5687	11/1924	8/1935	
688	4/1894	Gorton	12/1916	5688	2/1926	4/1931	
689	5/1894	Gorton	8/1911	5689	4/1924	2/1936	
690	5/1894	Gorton	6/1910	5690	12/1926	5/1937	
691	7/1894	Gorton	6/1918	5691	7/1926	7/1933	
692	8/1894	Gorton	3/1915	5692	11/1925	4/1937	
693	9/1894	Gorton	5/1912	5693	12/1925	2/1930	
700	10/1892	Kitson & Co	6/1913	5700	3/1926	5/1936	Westinghouse brake (1902-1932)
701	10/1892	Kitson & Co	11/1912	5701	10/1926	11/1935	
702	10/1892	Kitson & Co	7/1909	5702	5/1924	9/1933	
703	10/1892	Kitson & Co	7/1909	5703	8/1925	5/1935	
704	11/1892	Kitson & Co	1/1913	5704	1/1925	12/1939	
705	11/1892	Kitson & Co	2/1911	5705	9/1926	8/1933	Westinghouse brake from 1903
706	11/1892	Kitson & Co	12/1911	5706	8/1925	8/1935	
707	11/1892	Kitson & Co	6/1909	5707	4/1925	5/1933	
708	11/1892	Kitson & Co	10/1915	5708	4/1926	5/1937	
709	11/1892	Kitson & Co	1/1911	5709	3/1925	3/1932	
710	12/1892	Kitson & Co	1/1912	5710	3/1924	1/1930	
711	12/1892	Kitson & Co	12/1913	5711	2/1924	6/1933	

GC class '6DB' (D8)
Dimensions – see page 65

Statistics

MS&LR No.	Built	Builder	Reboilered	GC No.	LNER No.	Withdrawn
37	7/1888	Gorton	5/1912	508 (508B 3/20)	-	7/1923
89	10/1888	Gorton	5/1910	510 (510B 6/20)	6415 12/24	3/1926
400	10/1888	Gorton	12/1910	511 (511B 12/22)	-	11/1923

GC class '11B, C & D', (D9)
Dimensions – see page 67

Statistics

No.	Built	Builder	Rebuilt '11D'	LNER No.		1946 No.		BR No.		Withdrawn
1013	10/1901	Sharp Stewart	9/1918	6013	3/25	2300	12/46	62300	3/48	11/1949
1014*	10/1901	Sharp Stewart	12/1919	6014	10/24	2301	11/46	62301	7/48	4/1950
1015	10/1901	Sharp Stewart	6/1913	6015	7/24	2302	11/46	62302	9/48	3/1950
1016	10/1901	Sharp Stewart	11/1916	6016	7/25	2303	6/46	62303	2/49	8/1949
1017	10/1901	Sharp Stewart	4/1921	6017	2/24	2304	7/46	62304	2/49	1/1950
1018	2/1902	Sharp Stewart	8/1919	6018	1/26	2305	9/46	62305	12/48	7/1950
1019	2/1902	Sharp Stewart	11/1915	6019	7/25	2306	7/46	(62306)	1/1949	
1020	2/1902	Sharp Stewart	3/1914	6020	12/24	-		-		2/1943
1021*	2/1902	Sharp Stewart	4/1913	6021	2/25	2307	7/46	62307	9/48	6/1950
1022	2/1902	Sharp Stewart	6/1917	6022	10/24	-		-		8/1942
1023	2/1902	Sharp Stewart	5/1913	6023	11/25	2308	7/46	62308	6/48	8/1949
1024	2/1902	Sharp Stewart	4/1915	6024	2/24	2309	7/46	62309	2/49	11/1949
1025	3/1902	Sharp Stewart	8/1914	6025	3/25	(2310)		-		11/1945
1026	3/1902	Sharp Stewart	10/1914**	6026	1/25	2311	6/46	62311	3/49	7/1949
1027	3/1902	Sharp Stewart	7/1920**	6027	2/25	2312	6/46	62312	10/48	4/1950
1028	3/1902	Sharp Stewart	12/1914	6028	10/24	-		-		5/1939
1029	3/1902	Sharp Stewart	10/1913	6029	11/24	2313	7/46	62313	3/48	10/1949
1030	3/1902	Sharp Stewart	11/1915	6030	11/24	2314	7/46	62314	2/49	5/1949
1031	3/1902	Sharp Stewart	4/1922**	6031	8/24	2315	9/46	62315	6/48	7/1949
1032	3/1902	Sharp Stewart	6/1921	6032	2/24	(2316)		-		3/1945

No.	Built	Builder	Rebuilt '11D'	LNER No.		1946 No.		BR No.		Withdrawn
1033	3/1902	Sharp Stewart	1/1916	6033	6/24	2317	7/46	62317	9/48	7/1949
1034	3/1902	Sharp Stewart	8/1914	6034	3/24	2318	7/46	62318	2/49	11/1949
1035	4/1902	Sharp Stewart	4/1914	6035	6/25	2319	7/46	62319	3/49	7/1949
1036	4/1902	Sharp Stewart	5/1919	6036	7/25	(2320)		-		1/1946
1037	5/1902	Sharp Stewart	5/1916	6037	4/26	2321	7/46	62321	3/49	10/1949
1038	3/1903	Sharp Stewart	10/1914	6038	12/25	2322	9/46	(62322)		1/1949
1039	3/1903	Sharp Stewart	5/1922	6039	8/24	(2323)		-		1/1946
1040	3/1903	Sharp Stewart	3/1919	6040	1/25	2324	7/46	62324	3/49	11/1949
1041	3/1903	Sharp Stewart	6/1913	6041	4/24	2325	6/46	62325	6/48	2/1950
1042	3/1903	Sharp Stewart	1/1927**	6042	9/24	-		-		7/1939
104*	3/1904	Vulcan F'dry	6/1923+	5104	10/25	(2326)		-		9/1944
105	3/1904	Vulcan F'dry	11/1923	5105	12/25	(2327)		-		6/1945
106	4/1904	Vulcan F'dry	6/1914	5106	7/25	2328	8/46	-		3/1947
107	4/1904	Vulcan F'dry	9/1924	5107	9/24	2329	8/46	(62329)		2/1949
108	4/1904	Vulcan F'dry	10/1913	5108	4/26	2330	8/46	62330	1/49	8/1949
109	4/1904	Vulcan F'dry	5/1914	5109	8/24	2331	1/46	-		9/1946
110*	5/1904	Vulcan F'dry	9/1923+	5110	1/26	-		-		3/1942
111	5/1904	Vulcan F'dry	11/1916	5111	12/25	2332	8/46	62332	9/48	9/1949
112	5/1904	Vulcan F'dry	5/1924**	5112	5/24	2333	1/46	62333	8/48	12/1949
113	6/1904	Vulcan F'dry	5/1923**+	5113	6/26	-		-		6/1939

* 1014 named *Sir Alexander* 1902 - 1913
 1021 named *Queen Mary* 1913 – 1950
 104 named *Queen Alexandra* 1907 – 1944
 110 named *King George V* 1911 - 1942

** Rebuilt with new piston valves and cylinder block
   1026 12/1909
   1027 3/1918
   1031 5/1917
   1042 11/1922
   112 10/1918
   113 10/1918

+ 104 Fitted with large boiler as Class 11C, 1907 – 1923
  110 Fitted with large boiler as Class 11C, 1907 – 1918
  113 Fitted with large boiler as Class 11C, 1918 - 1923

GC class '11E', (D10)
Dimensions – see page 90

Statistics

No.	Name	Built	Builder	LNER No.		1946 No.		BR No.		Withdrawn
429	Prince Henry*	8/1913	Gorton	5429	8/24	2650	11/46	62650	3/49	2/1954
430	Purdon Viccars	9/1913	Gorton	5430	9/25	2651	10/46	62651	8/48	3/1953
431	Edwin A Beazley	10/1913	Gorton	5431	8/24	2652	10/46	62652	2/49	5/1954
432	Sir Edward Fraser	10/1913	Gorton	5432	7/24	2653	11/48	62653	11/48	10/1955
433	Walter Burgh Gair	10/1913	Gorton	5433	4/25	2654	9/46	62654	12/49	8/1953
434	The Earl of Kerry	11/1913	Gorton	5434	1/25	2655	3/49	62655	3/49	8/1953
435	Sir Clement Royds	11/1913	Gorton	5435	4/25	2656	8/46	62656	5/48	1/1955
436	Sir Berkeley Sheffield	11/1913	Gorton	5436	6/25	2657	11/46	62657	7/49	3/1953
437	Prince George**	11/1913	Gorton	5437	11/24	2658	8/48	62658	8/48	8/1955
438	Worsley Taylor	12/1913	Gorton	5438	10/24	2659	10/46	62659	1/50	11/1954

* Named *Sir Alexander Henderson* until 1917 & then *Sir Douglas Haig* until 1920.
** Named *Charles Stuart Wortley* until 1920.

GC class '11F', (D11)
Dimensions – see page 113

Statistics

No.	Name	Builder	Built	LNER No.		1946 No.		BR No.		Withdrawn
506	Butler-Henderson	Gorton	12/1919	5506	6/24	2660	10/46	62660	10/49	10/1960 P'served
507	Gerard Powys Dewhurst	Gorton	2/1920	5507	3/24	2661	10/46	62661	12/48	11/1960
508	Prince of Wales	Gorton	3/1920	5508	4/24	2662	10/46	62662	5/49	8/1960
509	Prince Albert	Gorton	3/1920	5509	8/24	2663	10/46	62663	8/49	5/1960
510	Princess Mary	Gorton	3/1920	5510	8/25	2664	9/46	62664	4/48	8/1960
501	Mons	Gorton	9/1922	5501	11/24	2665	5/46	62665	6/48	5/1959
502	Zeebrugge	Gorton	10/1922	5502	11/24	2666	7/46	62666	9/49	12/1960
503	Somme	Gorton	11/1922	5503	12/24	2667	7/46	62667	10/49	8/1960
504	Jutland	Gorton	11/1922	5504	12/24	2668	10/46	62668	5/48	11/1960
505	Ypres	Gorton	12/1922	5505	11/24	2669	10/46	62669	2/49	8/1960
511	Marne	Gorton	12/1922	5511	2/25	2670	10/46	62670	4/49	11/1960

LNER D11/2

No.	Name	Builder	Built	1946 No.		BR No.		Withdrawn
6378	Bailie MacWheeble	Kitson	7/1924	2671	9/46	62671	4/48	5/1961
6379	Baron of Bradwardine	Kitson	8/1924	2672	9/46	62672	9/51	9/1961
6380	Evan Dhu	Kitson	8/1924	2673	9/46	62673	11/48	7/1959
6381	Flora MacIvor	Kitson	8/1924	2674	4/49	62674	4/49	7/1961
6382	Colonel Gardiner	Kitson	8/1924	2675	5/49	62675	5/49	10/1959
6383	Jonathan Oldbuck	Kitson	8/1924	2676	11/46	62676	8/48	10/1959
6384	Edie Ochiltree	Kitson	9/1924	2677	8/46	62677	4/48	8/1959
6385	Luckie Mucklebackit	Kitson	9/1924	2678	8/46	62678	5/49	3/1959
6386	Lord Glenallan	Kitson	10/1924	2679	8/46	62679	11/50	9/1958
6387	Lucy Ashton	Kitson	10/1924	2680	6/46	62680	2/49	9/1961
6388	Captain Craigengelt	Kitson	10/1924	2681	9/46	62681	8/49	7/1961
6389	Haystoun of Bucklaw	Kitson	10/1924	2682	9/46	62682	6/49	7/1961
6390	Hobbie Elliott	Armstrong*	10/24	2683	9/46	62683	5/48	9/1958
6391	Wizard of the Moor	Armstrong*	10/24	2684	9/46	62684	2/50	10/1959
6392	Malcolm Graeme	Armstrong*	10/24	2685	8/46	62685	5/49	1/1962
6393	The Fiery Cross	Armstrong*	10/24	2686	10/46	62686	7/48	7/1961
6394	Lord James of Douglas	Armstrong*	10/24	2687	6/46	62687	9/49	8/1961
6395	Ellen Douglas	Armstrong*	11/24	2688	9/46	62688	9/48	7/1961
6396	Maid of Lorn	Armstrong*	11/24	2689	9/46	62689	7/49	7/1961
6397	The Lady of the Lake	Armstrong*	11/24	2690	8/46	62690	7/48	7/1961
6398	Laird of Balmawhapple	Armstrong*	11/24	2691	12/46	62691	1/50	11/1961
6399	Allan-Bane	Armstrong*	11/24	2692	1/47	62692	1/49	11/1959
6400	Roderick Dhu	Armstrong*	11/24	2693	11/46	62693	5/49	11/1961
6401	James Fitzjames	Armstrong*	11/24	2694	2/50	62694	2/50	11/1959

* Armstrong & Whitworth

GC class '6B', (D12)
Dimensions – see page 157

Statistics

No.	Built	Builder	Duplicate No.	LNER No.	Withdrawn
423	5/1877	Gorton	423 B	-	5/1923
424	6/1877	Gorton	424 B	-	1919
425	6/1877	Gorton	425 B	(6468)	7/1925
426	7/1877	Gorton	426 B	-	1919
427	7/1877	Gorton	427 B	-	1919
428	8/1877	Gorton	428 B	6467	3/1926
429	8/1877	Gorton	429 B	-	1922
430	9/1877	Gorton	430 B	6466	10/1926
431	9/1877	Gorton	431 B	-	1922
432	10/1877	Gorton	432 B	-	1921
433	11/1877	Gorton	433 B	-	1920
434	12/1877	Gorton	434 B	-	1/1923
435	1878	Gorton	435 B	-	1920
436	1878	Gorton	436 B	-	1921
437	1878	Gorton	437 B	-	1920
438	1878	Gorton	438 B	-	1922
439	5/1878	Gorton	439 B	(6465)	10/1925
440	6/1878	Gorton	440 B	-	5/1923
441	7/1878	Gorton	441 B	-	1/1923
442	10/1878	Gorton	442 B	6464	3/1930
443	11/1878	Gorton	443 B	(6463)	2/1925
444	12/1878	Gorton	444 B	-	1922
445	5/1879	Gorton	445 B	-	1920
446	6/1879	Gorton	446 B	-	4/1923
4	6/1880	Gorton	4 B	-	1920
128	7/1880	Gorton	128 B	6460	2/1926
129	8/1880	Gorton	129 B	-	1920

GE class 'T19 RBT', (LNER D13)
Dimensions – see page 164

All were built at Stratford as T19 2-4-0s and rebuilt there as 4-4-0s.

Statistics

No.	Built	Rebuilt as 4-4-0	25″ stroke	Superheated	LNER No.	Withdrawn
700	5/1892	3/1905		3/1920	7700	10/1935
704	5/1892	2/1906*	yes	6/1914	7704	12/1932
705	6/1892	10/1907	yes	-	7705	12/1926
706	6/1892	3/1905	yes	6/1925	7706	6/1938
707	6/1892	1/1905*		11/1913	7707	5/1937
708	6/1892	4/1906	yes	11/1917	7708	11/1935
710	11/1886	1/1908		10/1916	7710	12/1929
712	5/1887	5/1906	yes	11/1925	7712	11/1931
713	5/1887	1/1907		3/1922	7713	5/1933
717	6/1887	1/1907		-	7717	3/1930
718	6/1887	3/1905*	yes	3/1914	7718	5/1933
719	6/1887	5/1906*		3/1914	7719	3/1932
728	4/1888	5/1906*	yes	4/1914	7728	4/1931
729	4/1888	2/1905		2/1915	7729	4/1936
730	5/1888	2/1907		-	(7730)	8/1923
731	5/1888	2/1907	yes	1/1915	7731	8/1931
732	5/1888	4/1908	yes	5/1921	7732	5/1933
733	5/1888	7/1908	yes	12/1920	7733	9/1931
734	5/1888	6/1907		9/1919	7734	10/1931
735	6/1888	10/1907*		12/1913	7735	4/1932
737	6/1888	6/1907		3/1916	7737	4/1933
738	6/1888	2/1908		2/1921	7738	7/1931
739	6/1888	10/1907		6/1916	7739	9/1931
741	4/1889	6/1907		2/1920	7741	12/1935
742	4/1889	2/1907	yes	11/1914	7742	11/1935
744	5/1889	10/1907		8/1915	7744	4/1935
745	5/1889	6/1906		3/1918	7745	1/1933

No.	Built	Rebuilt as 4-4-0	25" stroke	Superheated	LNER No.	Withdrawn
748	5/1889	2/1905*		5/1914	7748	9/1931
751	6/1889	6/1907*		5/1914	7751	1/1933
756	9/1889	6/1908		6/1915	7756	12/1938
765	6/1890	3/1905	yes	3/1917	7765	1/1930
766	6/1890	2/1908		8/1919	7766	1/1935
767	6/1890	1/1908		2/1923	7767	12/1934
772	6/1892	2/1906	yes	10/1914	7772	9/1938
775	6/1892	2/1905*		11/1913	7775	5/1936
777	6/1892	3/1906*	yes	5/1914	7777	6/1934
779	7/1892	5/1906	yes	5/1918	7779	4/1935
1012	6/1893	5/1906	yes	4/1915	8012	12/1935
1013	6/1893	4/1906		6/1922	8013	10/1935
1015	6/1893	5/1906		11/1916	8015	10/1934
1016	7/1893	2/1906*		4/1914	8016	5/1938
1018	7/1893	7/1907		-	(8018)	6/1925
1020	6/1895	3/1907		12/1914	8020	4/1936
1021	6/1895	3/1906		10/1919	8021	8/1936
1023	7/1895	5/1906*		4/1914	8023	1/1944
1025	7/1895	3/1906		10/1914	8025	10/1937
1026	8/1895	4/1906		5/1926	8026	2/1937
1027	8/1895	5/1906		3/1915	8027	1/1936
1028	8/1895	3/1907		5/1918	8028	1/1938
1029	9/1895	3/1907		10/1919	8029	3/1938
1030	3/1897	4/1906		9/1916	8030	11/1938
1031	3/1897	3/1908		-	8031	12/1926
1032	3/1897	5/1906	yes	3/1921	8032	9/1936
1033	3/1897	3/1905		-	8033	6/1927
1035	4/1897	1/1905		3/1918	8035	5/1943
1036	4/1897	2/1907	yes	4/1918	8036	2/1937
1037	4/1897	3/1908		12/1920	8037	12/1934
1039	5/1897	10/1907		12/1921	8039	3/1944

* with stepped frame in front of smokebox

GE classes 'S46, D56 & H88', (LNER D14, D15 & D16)
Dimensions – see pages 171 & 173

All were built at Stratford.

Statistics
1860 – 1900 built as S46 (LNER D14)
1790 – 1859 built as D56 (LNER D15)
1780 – 1789 built as H88 (LNER D16/1)

No.	Built	Superheated	LNER No.	Rebuilt as D15	D15/2	D16/1	D16/2	D16/3	1946 No.	BR No.	Withdrawn
1900 *Claud Hamilton*	3/1900	3/1925	8900 3/25	1/29				2/33*	2500		5/1947
1890	4/1900	2/1916	8890	2/16	4/29				2501	62501	6/1951
1891	4/1900	3/1916	8891	3/16	6/32				2502	62502	2/1952
1892	4/1900	10/1927	8892	10/27	5/32				2503	62503	2/1951
1893	5/1900	12/1926	8893	12/26	5/31				2504		6/1948
1894	5/1900	1/1918	8894 +	1/18	4/32				2505	62505	11/1951
1895	6/1900	11/1926	8895	11/26	2/30				2506	62506	4/1952
1896	6/1900	8/1915	8895	8/15	6/33				2507	62507	4/1952
1897	6/1900	9/1919	8896	9/19	5/29				2508	62508	10/1950
1898	7/1900	6/1915	8897	6/15	4/33				2509	62509	9/1952
1899	7/1900	6/1925	8898	2/22	7/29			9/43	2510	62510	10/1957
1880	4/1901	2/1918	8880	2/18	3/28			9/42	2511	62511	12/1959
1881	5/1901		8881		5/30				2512	62512	8/1950
1882	5/1901	12/1919	8882	12/19	5/33			9/42	2513	62513	11/1958
1883	5/1901	1/1916	8883	1/16	8/28			5/43	2514	62514	3/1957
1884	5/1901	12/1928	8884	3/22	12/28			1/43	2515	62515	4/1958
1885	5/1901	5/1919	8885	5/19	12/28			7/37	2516	62516	8/1957
1886	6/1901	3/1929	8886	10/25	3/29			4/40	2517	62517	9/1959
1887	6/1901	2/1918	8887	2/18	4/33			8/43	2518	62518	10/1958
1888	6/1901	11/1918	8888	11/18	2/29			6/39	2519	62519	1/1957
1889	6/1901	3/1917	8889	3/17	2/30				2520	62520	8/1951
1870	3/1902	4/1924	8870+	4/24	6/28			5/35	2521	62521	2/1958
1871	3/1902	4/1924	8871	4/24	7/29			5/38	2522	62522	8/1958
1872	3/1902	3/1924	8872	3/24	4/28			5/40	2523	62523	8/1956

				Rebuilt as							
No.	Built	Superheated	LNER No.	D15	D15/2	D16/1	D16/2	D16/3	1946 No.	BR No.	Withdrawn
1873	3/1902	3/1933	8873	1/30	3/33			6/39	2524	62524	3/1960
1874	3/1902	5/1919	8874	5/19	12/32			6/38	2525	62525	9/1955
1875	4/1902	3/1931	8875		3/31			8/37	2526	62526	5/1957
1876	4/1902	6/1929	8876		6/29			3/35	2527	62527	7/1952
1877	4/1902	2/1928	8877		2/28				2528	62528	6/1951
1878	5/1902	7/1931	8878	3/29	7/31			5/35	2529	62529	11/1959
1879	5/1902	12/1927	8879		12/27			4/38	2530	62530	9/1958
1860	5/1903	5/1919	8860	5/19	6/28			6/34	2531	62531	3/1955
1861	5/1903	11/1929	8861	11/21	11/29			4/36**	2532	62532	11/1956
1862	5/1903	12/1923	8862	12/23	5/30			6/40	2533	62533	9/1957
1863	6/1903	6/1926	8863	11/21	6/26			4/35	2534	62534	11/1958
1864	6/1903	5/1923	8864	5/23	3/32		10/36**		2535	62535	11/1957
1865	9/1903	4/1930	8865		4/30			7/36**	2536	62536	7/1955
1866	9/1903	6/1923	8866	5/23	6/28			3/33*			9/1945
1867	10/1903	5/1923	8867	5/23	12/33				2538	62538	4/1952
1868	10/1903	9/1929	8868		9/29			7/40	2539	62539	10/1957
1869	11/1903	2/1924	8869	1/22	7/28			6/34	2540	62540	8/1959
1850	12/1903	12/1921	8850		2/32			7/39	2541	62541	10/1955
1851	12/1903	11/1918	8851				6/27	6/38	2542	62542	10/1956
1852	12/1903	7/1922	8852				2/28	2/49	2543	62543	10/1958
1853	12/1903		8853			5/26	11/29	3/47	2544	62544	11/1959
1854	1/1904	6/1916	8854					3/33	2545	62545	9/1958
1855	2/1904	11/1914	8855		9/28			1/34*	2546	62546	6/1957

Claud Hamilton from 8/47

No.	Built	Superheated	LNER No.	D15	D15/2	D16/1	D16/2	D16/3	1946 No.	BR No.	Withdrawn
1856	3/1904	9/1923	8856					5/27	2547	62547	2/1951
1857	3/1904	11/1916	8857+		10/33			7/39	2548	62548	10/1957
1858	3/1904	5/1927	8858+		5/30			3/38	2549	62549	12/1955
1859	4/1904	5/1931	8859		5/31			5/34	2550		11/1946
1840	11/1906	8/1923	8840		2/29			3/35	2551	62551	7/1956
1841	11/1906	1/1929	8841				1/29	2/49	2552	62552	10/1955
1842	11/1906	9/1930	8842+				9/30	9/49	2553	62553	1/1957
1843	11/1906	7/1927	8843				7/27	11/38	2554	62554	11/1955
1844	11/1906	10/1914	8844		11/28			3/39	2555	62555	3/1958
1845	11/1906	10/1914	8845				5/29	4/46	2556	62556	1/1957

| | | | | Rebuilt as | | | | | | | |
No.	Built	Superheated	LNER No.	D15	D15/2	D16/1	D16/2	D16/3	1946 No.	BR No.	Withdrawn
1846	12/1906	12/1924	8846			12/24	3/34	4/44	2557	62557	10/1955
1847	12/1906	5/1921	8847			7/26	10/33	9/48	2558	62558	5/1957
1848	1/1907	9/1918	8848		9/29			1/33	2559	62559	12/1955
1849	1/1907	5/1919	8849		5/28			4/33*	2560		9/1948
1830	3/1908	1/1925	8830		5/29			3/40	2561	62561	2/1958
1831	3/1908	4/1916	8831				4/28	2/46	2562	62562	10/1957
1832	3/1908	7/1919	8832		11/30			7/36**	2563		8/1948
1833	4/1908	1/1930	8833				1/30	1/48	2564	62564	3/1958
1834	4/1908	3/1929	8834				3/29	6/45	2565	62565	1/1957
1835	5/1908	3/1925	8835		5/29			1/39	2566	62566	12/1958
1836	5/1908	6/1928	8836		2/30			10/37	2567	62567	12/1956
1837	6/1908	7/1931	8837+		7/31			5/33*	2568	62568	4/1958
1838	7/1908	11/1929	8838+				11/29	5/48	2569	62569	11/1956
1839	7/1908	2/1922	8839				4/28	9/49	2570	62570	11/1959
1820	6/1909	5/1932	8820		5/32			5/39	2571	62571	1/1959
1821	6/1909	5/1920	8821					5/33	2572	62572	7/1958
1822	9/1909	6/1930	8822+				6/30	3/47	2573	62573	10/1955
1823	11/1909	12/1921	8823		12/29			4/38	2574	62574	12/1955
1824	11/1909	4/1918	8824		1/30			2/40	2575	62575	5/1957
1825	11/1909	5/1922	8825+		6/28			3/37**	2576	62576	9/1957
1826	12/1909	7/1922	8826				4/29	5/49	2577	62577	10/1956
1827	12/1909	7/1927	8827				7/27	9/44	2578	62578	10/1957
1828	12/1909	6/1928	8828		6/28			3/34	2579	62579	3/1955
1829	12/1909	2/1930	8829				2/30	4/48	2580	62580	6/1958
1810	3/1910	4/1922	8810		3/31			7/36**	2581	62581	3/1953
1811	3/1910	4/1922	8811		3/32			12/39	2582	62582	1/1959
1812	3/1910	3/1922	8812		4/34			6/36**	2583		11/1948
1813	4/1910	7/1914	8813				4/26	9/47	2584	62584	12/1957
1814	4/1910	11/1921	8814		12/31			2/35	2585	62585	4/1955
1815	5/1910	11/1922	8815		5/31			8/39	2586	62586	3/1958
1816	6/1910	5/1920	8816		12/29			3/34*	2587	62587	12/1956
1817	6/1910	5/1924	8817		5/30			7/34*	2588	62588	10/1958
1818	6/1910	6/1923	8818			6/23	11/33	3/47	2589	62589	5/1959
1819	6/1910	6/1919	8819		1/26		1/28		2590	62590	1/1952

| | | | | Rebuilt as | | | | | | | |
No.	Built	Superheated	LNER No.	D15	D15/2	D16/1	D16/2	D16/3	1946 No.	BR No.	Withdrawn
1800	7/1910	7/1927	8800				7/27		2591	62591	4/1950
1801	7/1910	6/1919	8801				4/29	6/45	2592	62592	4/1958
1802	8/1910	5/1928	8802		5/28			4/33	2593	62593	10/1957
1803	8/1910	9/1927	8803		9/27			4/37**	2594		3/1949
1804	8/1910	12/1928	8804		12/28			3/34*	2595		11/1946
1805	9/1910	3/1923	8805			3/23	11/31	3/47	2596	62596	10/1957
1806	9/1910	8/1929	8806		8/29			1/40	2597	62597	1/1960
1807	10/1910	6/1914	8807		12/29			1/42	2598	62598	5/1952
1808	10/1910	7/1929	8808		7/29			6/37**	2599	62599	9/1958
1809	11/1910	7/1929	8809					6/33**	2600		6/1948
1790	2/1911	8/1927	8790		8/27		4/29	9/44	2601	62601	1/1957
1791	3/1911	4/1923	8791		5/28			2/37**	2602		9/1948
1792	3/1911	7/1914	8792				3/28		2603	62603	9/1951
1793	4/1911	4/1911	8793		5/29			7/37	2604	62604	2/1960
1794	5/1911	5/1911	8794				2/29	3/40	2605	62605	6/1957
1795	7/1911	4/1931	8795+				4/31	3/46	2606	62606	9/1959
1796	7/1911	4/1925	8796				12/28	12/46	2607	62607	11/1955
1797	7/1911	7/1914	8797		4/28			7/37	2608	62608	1/1957
1798	8/1911	8/1911	8798		4/32			7/37	2609	62609	2/1957
1799	8/1911	8/1911	8799		2/34			4/34*	2610	62610	1/1959
1780	6/1923	6/1923	8780			6/23	1/28	9/44	2611	6261	1/1957
1781	6/1923	6/1923	8781			6/23	8/28	4/49	2612	62612	11/1959
1782	6/1923	6/1923	8782			6/23	5/31	12/48	2613	62613	10/1960
1783	7/1923	7/1923	8783			7/23	6/28	12/39	2614	62614	8/1958
1784	7/1923	7/1923	8784			7/23	6/29	4/47	2615	62615	10/1958
1785	7/1923	7/1923	8785			7/23	6/28	5/44	2616	62616	2/1953
1786	8/1923	8/1923	8786+			8/23	6/29	1/45	2617	62617	5/1957
1787	8/1923	8/1923	8787			8/23	1/29	8/44	2618	62618	11/1959
1788	9/1923	9/1923	8788			9/23	12/28	12/38	2619	62619	10/1957
1789	9/1923	9/1923	8789			9/23	12/28	3/48	2620	62620	10/1955

* 8" diameter Piston valves
** 9 ½" diameter piston valves
+ Numbered in 77XX series 1943-1946

M&GN 'A' Rebuilds
For Dimensions, see page 207

Statistics
All built by Beyer, Peacock.

M&GNR No.	Built	Reboilered	2nd Reboilered	LNER No.	Withdrawn
23	3/1882	1895	1919	-	2/1937
25	11/1883	1906	1920	025*	5/1941
26	11/1883	1904	1923	-	11/1936
27	11/1883	1905	1927	027	2/1937
28	11/1883	1905	1925	-	2/1938

* boiler of 25 on frames of 24 during works visit, 11/1936.

M&GN 'C' (LNER D52, D53 & D54)
For Dimensions, see pages 209 & 212

Statistics
01 – 057 built by Sharp, Stewart & Co., 074 – 080 built by Beyer, Peacock & Co, all as M&GN class 'C'. All were reboilered with Midland class boilers during their lifetime.

M& GNR No.	Built	LMS boiler	Rebuilt	Class	LNER No.	Withdrawn
1	8/1894	1932		D52	01	11/1937
2	8/1894		4/1931	D53	02	5/1943
3	8/1894	1932		D52		6/1937
4	8/1894	1933		D52		2/1938
5	8/1894	1935		D52	05	7/1937
6	8/1894		8/1930	D53	06	3/1944
7	8/1894	1933		D52	07	6/1937
11	9/1894	1933		D52	011	8/1942
12	11/1894	1932		D52	012	8/1942
13	11/1894			D52	013	9/1941
14	11/1894	1934		D52		2/1937
17	11/1894	1935		D52		10/1937
18	11/1894			D52		2/1937
36	5/1894		5/1929	D53		1/1937
37	5/1894			D52		2/1937

M& GNR No.	Built	LMS boiler	Rebuilt	Class	LNER No.	Withdrawn
38	5/1894	1931		D52	038	9/1943
39	5/1894	1908*	1/1924	D54		2/1937
42	5/1894	1931		D52	042	6/1940
43	5/1894			D52	043	6/1943
44	6/1894		5/1930	D53	044	8/1941
45	6/1894		1909	D54		11/1936
46	6/1894		1915	D54	046	3/1943
47	6/1894			D52	047	6/1942
48	7/1894			D52		11/1937
49	7/1894		2/1931	D53	049	9/1941
50	7/1894		11/1929	D53	050	1/1945
51	8/1896		1915	D54	051	5/1943
52	8/1896		1913	D54	052	2/1943
53	8/1896		1910	D54	053	1/1940
54	9/1896		1914	D54	054	10/1939
55	9/1896	1908*	7/1925	D54	055	11/1943
56	9/1896		1912	D54	056	11/1943
57	9/1896		1912	D54		2/1937
74	10/1899			D52		5/1937
75	10/1899			D52		2/1937
76	10/1899	1936		D52	076	7/1943
77	11/1899		12/1930	D53	077	1/1945
78	11/1899			D52	078	2/1938
79	11/1899			D52	079	2/1937
80	11/1899			D52		2/1937

* Midland 'H' boiler

BIBLIOGRAPHY

Aldrich, C.Langley, *The Locomotives of the Great Eastern Railway, 1862-1962,* The 'Langloco' series, 1969
Becket, W.S., *Xpress Locomotive Register, Vol.3 Eastern, North Eastern & Scottish (ex LNER) Regions, 1950-1960,* Xpress Publishing
Maidment, David J., *LNER 4-6-0 Locomotive Classes,* Pen & Sword, 2021
Maidment, David J., *Midland & LMS 4-4-0 Locomotive Classes,* Pen & Sword, 2021
RCTS, *Locomotives of the LNER Vol 3B Tender Engines – Classes D1 – D12,* RCTS, 1980
RCTS, *Locomotives of the LNER Vol 3C Tender Engines – Classes D13 – D24,* RCTS, 1981
Tuplin, W.A., *Great Central Steam,* George Allen & Unwin, 1967
Yeadon's *Register of LNER Locomotives Vol. 14 – Class D13 –D16,* Book Law Publications, 1999
Yeadon's *Register of LNER Locomotives Vol. 29 – Class D5 –D12,* Book Law Publications, 2003
Yeadon's *Register of LNER Locomotives Vol. – Class D1 –D4,* Book Law Publications, XXXX

INDEX

All references to LNER and pre-Grouping locomotives will use the LNER class identification and 1923-45 numbers (Pre-Grouping, 1946 LNER or BR numbers in brackets when photo is of that era).

4-4-0 Comparisons, Introduction, 140

Engineers
Hill, Alfred J., 10
Holden, James, 9
Holden, Stephen, 10
Ivatt, H.A., 8
Johnson, Samuel, 10
Parker, Thomas, 8-9
Pollitt, Henry, 9
Robinson, John G., 9
Sacré, Charles, 8

(Great Northern – summary), 11
D1
Allocation, 14, 16, 18
Braking systems, 12-13
Dimensions, 11
Livery, 13
Lubrication, 12
Numbers, 11-13
Operation, 14-18
Statistics, 225
Superheating, 12
Tenders, 13
Withdrawal, 13, 18

D2
Allocation, 23, 26, 28
Dimensions, 19
Livery, 21
Marshall's valve gear, 19
Numbers, 18-19, 21
Operation, 21-23
Performance, 27-28
Rebuilding, 20
Statistics, 225-227
Superheating, 20
Tenders, 21
Withdrawals, 28

D3 & D4
Allocation, 33, 36-38, 40-42
D4 unrebuilt, 33
Dimensions, 30
Livery, 32
Numbers, 30, 32-33
Operation, 36, 38, 40
Performance, 36
Rebuilding, 30, 32-33
Statistics, 228-229
Tenders, 30
Withdrawal, 33, 38, 42

(Great Central – summary), 43
D5
Construction, 43
Dimensions, 43-44
Numbers, 43
Operation, 44, 47
Performance, 47
Reboilering, 44
Statistics, 229
Superheating, 44
Withdrawal, 44, 47

D6
Allocation, 50-51, 54
Construction, 47
Dimensions, 47
Livery, 48
Lubrication, 48
Numbers, 47,
Operation, 50-51
Statistics, 230
Superheating, 47
Tenders, 48
Withdrawal, 48, 54

D7
Allocation, 63
Braking, 58
Construction, 56
Design, 56
Dimensions, 56
Livery, 58
Numbers, 58
Operation, 60
Performance, 63-64
Preservation, 64
Reboilering, 56
Statistics, 231
Withdrawal, 58, 63

D8
Allocation, 67
Construction, 64
Design, 64
Dimensions, 65

Numbers, 64
Operation, 67
Reboilering, 66
Statistics, 232
Withdrawal, 66-67

D9
Allocation, 77-78, 83
Construction, 67-68
Dimensions, 67-69
Livery, 69
Names, 68
Numbers, 67-69
Operation, 73
Performance, 77-78, 82-84
Reboilering, 68
Statistics, 232-233
Superheating, 68
Tenders, 68
Withdrawal, 69, 84

D10
Allocation, 95, 104, 106-107, 109, 111
Construction, 89
Design, 89
Dimensions, 90
Livery, 90, 92
Names, 90, 92
Numbers, 89, 92
Operation, 95
Performance, 99-101, 104-107, 109
Statistics, 234
Superheating, 90
Trofinoff by-pass valves, 92
Withdrawal, 92, 111

D11/1
Allocation, 127-128, 131, 133-134
Construction, 113
Dimensions, 113
Livery, 115
Long travel valves, 115
Names, 113
Numbers, 115-116
Operation, 116
Performance, 122, 124, 127-129, 131, 134

Preservation, 140-141
Special trains, 134
Statistics, 234
Withdrawal, 116, 134

D11/2
Allocation, 148-149, 153
Construction, 143
Dimensions, 143
Introduction, 142
Livery, 144
Names, 143
Numbers, 143-144
Operation, 149-150, 152
Performance, 149
Statistics, 235
Withdrawal, 144, 153

D12
Allocation, 159
Braking concerns, 157
Construction, 157
Dimensions, 157
Numbers, 157-158
Operation, 159
Statistics, 236
Withdrawal, 158

D13
Allocation, 168, 170
Design, 162
Dimensions, 164
Livery, 165
Numbers, 162, 164-165
Operation, 168
Performance, 169-170
Rebuilding T19 as D13, 164
Statistics, 237-238
Superheaters, 165
T19RBT 2-4-0, 162
Withdrawal, 165, 169-170

D14
Design, 170
Dimensions, 171-172
Livery, 172
Numbers, 170

Oil burning, 171
Paris Exhibition, 171
Statistics, 239-242
Rebuilding as D15, 172

D15
Construction, 173
Dimensions,, 173
Livery, 174, 176
Numbers, 173-174, 176
Rebuild of D14 with Belpaire Firebox, 173
Statistics, 239-242
Steaming problems, 176
Superheating, 173-174
Withdrawal, 176

D16
Allocation, 187, 191, 194, 197, 199-201
Construction, 179
Design, 179
Dimensions, 179
Livery, 179, 199
Numbers, 179, 182-183
Operation, 183, 187, 191, 194-195, 199
Performance, 187-188, 194, 196-198
Reboilering, 182
Rebuilt with new cylinders and piston valves, 182-183
Statistics, 239-242
Steaming problems resolution, 182
Withdrawal, 183, 199-201

M&GN 'A' class
Construction, 207
Dimensions, 207
Numbers, 207
Operation, 208
Rebuilt, 207
Statistics, 243
Withdrawal, 207-208

D52 (ex M&GN)
Allocation, 209
Construction, 209
Dimensions, 209

Numbers, 209
Statistics, 243-244
Withdrawal, 210

D53 & D54 (ex M&GN)
Allocation, 212
Dimensions, 212
Numbers, 212
Operation, 212
Performance, 212
Rebuilt C class, 212
Statistics, 243-244
Taken over by LNER, 212

Logs
Cambridge – Liverpool St., 195
Dundee – Edinburgh, 150
Glasgow – Edinburgh, 17, 148-149
Grantham – York, 15, 23
Harrogate – King's Cross, 123-124
High Wycombe – Woodford, 100
Ipswich - Norwich, 191
Leicester – Sheffield, 99
Liverpool St – North Walsham, 188
Manchester – Liverpool, 52-54, 80-81, 109-110
Manchester – Penistone, 101
Marylebone – Leicester, 97-98, 104-105, 120-121, 126
Sheffield – Marylebone, 78
Sheffield – Manchester, 106-107, 128

Models
62581 (Hornby), 206
62690 (Bachmann), 156

Photographs of Locations (B&W)
Aberdeen Ferryhill, 151
Allerton Sidings, 67
Annesley, 158-159
Arisaig, 155
Ashby, 85
Ashley, 55-56, 85, 110-111, 136-137
Barnsley Court House, 65
Basford, 37
Beccles, 193
Belle Isle, 25
Bishopbriggs, 154

Breadsall, 26
Brentwood, 197
Cambridge, 202
Carlisle Canal, 13
Charwelton, 102
Cheadle, 55, 85, 89
Chester Northgate, 96, 136
Clapham BTC Museum, 142
Colwick, 22
Cromer, 194
Darlington Centenary, 175
Dereham, 203-204
Dinting, 132
Doncaster, 20, 22, 25, 29, 32, 35-36
Dundee, 146
Dunsford Bridge, 129
Edinburgh Princes St Gdns, 152-153
Edinburgh Waverley, 155
Flixton, 63, 133, 138
Forth Bridge, 17
Glasgow Eastfield, 143, 145, 147
Gorton, 47, 60, 72, 91, 93, 114, 124, 141
Grantham, 35, 42, 58, 61, 94
Grimsby, 64
Grindley, 29
Guide Bridge, 130
Harringay, 24
Haymarket, 14
Hazlehead Bridge, 122
Hull Paragon, 39, 76
Hunstanton, 201
Hunts Cross, 76
Hyde Junction, 108
Immingham, 118
Inverkeithing, 151
Kings Cross, 19, 32
Kings Lynn, 178, 186
Kirkby Stephen, 39
Knutsford, 112
Leicester, 50, 73, 101, 158
Lincoln, 41, 46
Lincoln St Marks, 119, 135
Liverpool Brunswick, 71, 117
Longniddry, 148
Louth, 45
Luddenfoot, 135
Manchester Central, 49, 58, 71, 84, 87-88, 95, 112, 160

Manchester Exchange, 134
Manchester London Rd, 119, 131
March, 184, 186-187
Marshmoor, 126
Melton Constable, 202-203, 208
Neasden, 74, 114-115
New Holland, 59-60, 157
Newark, 27
New Holland, 33
Northwich, 96
Nottingham, 215
Nottingham Victoria, 28, 31, 75, 94
Parkend Colliery, 108
Penistone, 61, 107
Perth, 154
Peterborough East, 198
Potters Bar, 196
Prioy Junction, 142
Retford, 48, 82, 137
Risley Moss, 200
Sefton, 86
Sheffield Darnall, 117
Sheffield Midland, 118
Sheffield Victoria, 79, 103, 139-140, 205
Shenfield, 189-190
Silloth, 18
Skelton Junction, 86
Southport, 66, 201
Spalding, 34, 208
St Pancras, 171
Staveley, 139
Stratford, 167, 173, 175-176, 180, 183
Tebay, 40
Thornton Junction, 146, 147
Thurmaston, 215
Trafford Park, 45, 48, 161
Vale House, 132
Waleswood, 138
Waltham Cross, 164
Wavertree Junction, 87
Weybourne, 205
Whetstone, 75
Wisbech, 204
Wood Green, 126
Woodhead, 62
Yarmouth Beach, 211, 216
Yarmouth South Town, 177, 185

Photographs of Locations (Colour)
Bishops Stortford, 222
Caister, 224
Calder Bridge Junction, 219
Cambridge, 223-224
Grantham, 217
Gunton, 223
Hare Park, 219
Haymarket, 220
Heacham, 221
Leicester, 223
Longniddry, 220
Manchester Central, 217
Nottingham Victoria, 218-220
Sheffield Darnall, 218
Sudbury, 222
Trent Junction, 218
Trowse Swing Bridge, 221
Welwyn North, 221
Wintersett Junction, 217

Photographs of Locomotives (B&W)
O1 (D52), 211
O6 (D53), 216
14 (D52), 210
18 (D52), 210
24 (M&GN A), 207
26 (M&GN A), 208
28 (M&GN A), 208, 210
45 (D54), 213
O47 (D52), 211
O52 (D54), 216
53 (D54), 215
54 (D54), 214
56 (D54), 214
77 (D53), 212-213, 215
400 (511B), 66
424 (D12), 160
431 (D12), 160
438 (D12), 157
702 (T19), 164
760 (T19), 163
763 (T19), 163
784 (T19), 162
3042, 21
3044, 26
3050 (50), 24
3057, 17
3058 (2209), 13, 14, 18
3059, 13, 16
3064, 14
3251 (251 C1), 134
3400 (400), 37, 32
4075 (2000), 34, 35, 40, 42
4077 (1077), 30
4080 (1080), 32
4306, 37
4309 (2126), 36
4311, 38
4313, 39
4316, 33
4321, 20
4326 (1326), 19
4333 (2161), 28
4335 (2162), 22
4336 (1336), 25
4343 (2135), 41
4348 (1348), 31, 39
4351, 34
4354 (1354), 31
4358 (1358), 31
4369 (62172), 22, 29
4370 (1370), 19
4375, 27
4381 (1381), 24
4386 (1386), 25
4395 (2190), 29
4396, 20
4406 (1406) C1, 25
4460 (C1), 38
5104 (104), 69, 71, 74
5106, 82
5110, 70
5111 (111, 2322, 62332), 73, 85, 86
5112 (112, 62333), 74, 89
5113 (113), 68-69
5429 (429, 62650), 90, 95, 102
5430 (430), 91, 108, 112
5431 (62652), 96
5432 (62653), 96
5434 (2655, 62655), 107, 110-112
5435, (62656) 94-95
5436 (436), 91, 103
5437 (437), 92-93, 102-103
5438 (62659), 94, 108, 112
5501 (62665),115-116, 131, 138
5502 (502), 115, 117, 122, 125, 132
5503 (62667), 118-119, 135
5504 (504), 114
5505 (62669), 117, 130
5506 (506, 62660), 113, 125, 140-142
5507 (62661), 137
5508 (62662), 132, 135-136, 138-139
5509 (2663, 62663), 118, 134
5510 (510, 62664), 114, 129, 135-136
5511 (62670), 124, 130, 133, 139
5561 (561), 57
5566 (566), 58
5567, 65
5684, 59
5685 (685), 58, 61
5687, 60, 64
5689 (680), 61
5694 (694), 44-46
5697, 45
5699 (699), 46
5703 (703), 62
5704, 60
5705 (705), 59
5707 (707), 63
5708 (708), 62
5855 (855), 52, 55
5856 (856), 47
5859 (859), 48
5869, 49
5871, 49, 55
5874 (2106), 56
5880, 48
5881 (881), 51
6013 (62300), 84, 88
6014 (1014), 76
6015 (62302), 72, 85
6016, 79
6018 (2305, 62305), 85, 88
6021 (1021), 69
6027 (62312), 87
6024 (1024), 75
6029 (1029, 62313), 75, 87
6033 (62317), 72
6034 (2318), 71
6035 (2319), 84
6037 (1037), 79
6038 (2322), 85
6040 (1040), 76

6379, 143
6380 (62673), 147
6381 (62674), 154
6382, 145
6384 (62677), 156
6385 (62678), 144, 147, 153
6386 (62679), 152, 155
6388, 145
6389 (62682), 151, 155
6393 (62686), 147
6394, 144, 151
6396 (62689), 148
6398 (62691), 154
6399 (62692), 146
6400 (62693), 148
6401 (2694), 146
6415 (510B), 66-67
6463 (443), 161
6464 (442), 158
6466 (438), 157, 158
7707, 166
7730 (730), 169
7751, 166
7791 (1791), 175
7904 (J15), 170
8020, 167
8025 (1025), 165
8026, 166
8032, 170
8035 (1035), 165, 168

8044 (F3), 166
8781 (62612), 180, 202
8783, 196, 206
8785 (1785), 179
8786, 192
8787, 181, 186-187, 197
8789 (62620), 202
8801 (62592), 204
8812, 184
8816 (1816), 189
8820 (62571), 205
8826 (62577), 203
8827 (62578), 203
8828 (1828), 174, 176
8830 (62561), 205
8831 (62562), 204
8838, 193
8839, 180
8842 (1842), 189, 200
8855 (62546), 185
8858 (1858), 190
8857 (2548), 184
8864, 200
8865 (62536), 201
8870 (62521), 185
8872 (1872), 172
8873, 193-194
8876, 198
8879, 173, 186
8891 (2502), 177-178

8894 (2505), 177
8896 (62507), 178, 181, 201
8900 (1900), 171, 175, 183, 192
41235 (2MT), 112
62787 (E4), 204
65179 (J10), 86
67154 (F5), 185

Photographs of Locomotives (Colour)
4075 (2000), 217
5435 (62656), 217
5501 (62665), 218
5503 (62667), 217-218
5504 (62668), 218
5505 (62669), 219-220
5506 (62660), 218-219
6380 (62673), 220
6384 (62677), 220
8783 (62614), 221, 224
8787 (62618), 221, 224
8805 (62596), 223
8820 (62571), 223
8822 (2573), 222
8823 (62574), 222
8879 (62530), 223
8886 (62517), 224
8893, 222
D13 (unknown), 221